NATIONAL ACADEMIES *Sciences Engineering Medicine*

NATIONAL ACADEMIES PRESS
Washington, DC

Navigating an Electric Vehicle Future

Raphael Apeaning and Kyra Howe,
Rapporteurs

Board on Energy and Environmental Systems

Division on Engineering and Physical Sciences

Proceedings of a Workshop

THE NATIONAL ACADEMIES PRESS 500 Fifth Street, NW Washington, DC 20001

This activity was supported by Award No. DTNH2217H00028 between the National Academy of Sciences and the National Highway Traffic Safety Administration of the U.S. Department of Transportation. Any opinions, findings, conclusions, or recommendations expressed in this publication do not necessarily reflect the views of any organization or agency that provided support for the project.

International Standard Book Number-13: 978-0-309-69219-9
International Standard Book Number-10: 0-309-69219-9
Digital Object Identifier: https://doi.org/10.17226/26668

This publication is available from the National Academies Press, 500 Fifth Street, NW, Keck 360, Washington, DC 20001; (800) 624-6242 or (202) 334-3313; http://www.nap.edu.

Printed in the United States of America.

Suggested citation: National Academies of Sciences, Engineering, and Medicine. 2022. *Navigating an Electric Vehicle Future: Proceedings of a Workshop*. Washington, DC: The National Academies Press. https://doi.org/10.17226/26668.

The **National Academy of Sciences** was established in 1863 by an Act of Congress, signed by President Lincoln, as a private, nongovernmental institution to advise the nation on issues related to science and technology. Members are elected by their peers for outstanding contributions to research. Dr. Marcia McNutt is president.

The **National Academy of Engineering** was established in 1964 under the charter of the National Academy of Sciences to bring the practices of engineering to advising the nation. Members are elected by their peers for extraordinary contributions to engineering. Dr. John L. Anderson is president.

The **National Academy of Medicine** (formerly the Institute of Medicine) was established in 1970 under the charter of the National Academy of Sciences to advise the nation on medical and health issues. Members are elected by their peers for distinguished contributions to medicine and health. Dr. Victor J. Dzau is president.

The three Academies work together as the **National Academies of Sciences, Engineering, and Medicine** to provide independent, objective analysis and advice to the nation and conduct other activities to solve complex problems and inform public policy decisions. The National Academies also encourage education and research, recognize outstanding contributions to knowledge, and increase public understanding in matters of science, engineering, and medicine.

Learn more about the National Academies of Sciences, Engineering, and Medicine at **www.nationalacademies.org**.

Consensus Study Reports published by the National Academies of Sciences, Engineering, and Medicine document the evidence-based consensus on the study's statement of task by an authoring committee of experts. Reports typically include findings, conclusions, and recommendations based on information gathered by the committee and the committee's deliberations. Each report has been subjected to a rigorous and independent peer-review process and it represents the position of the National Academies on the statement of task.

Proceedings published by the National Academies of Sciences, Engineering, and Medicine chronicle the presentations and discussions at a workshop, symposium, or other event convened by the National Academies. The statements and opinions contained in proceedings are those of the participants and are not endorsed by other participants, the planning committee, or the National Academies.

Rapid Expert Consultations published by the National Academies of Sciences, Engineering, and Medicine are authored by subject-matter experts on narrowly focused topics that can be supported by a body of evidence. The discussions contained in rapid expert consultations are considered those of the authors and do not contain policy recommendations. Rapid expert consultations are reviewed by the institution before release.

For information about other products and activities of the National Academies, please visit www.nationalacademies.org/about/whatwedo.

PLANNING COMMITTEE FOR NAVIGATING AN ELECTRIC VEHICLE FUTURE

GARY MARCHANT, Arizona State University, *Chair*
CARLA BAILO, Center for Automotive Research
NADY BOULES, NB Motors, LLC
ZHENHONG LIN, Oak Ridge National Laboratory
JOSHUA LINN, University of Maryland, College Park
DEIDRE STAND, Wildcat Discovery Technologies
ANNA STEFANOPOULOU, University of Michigan
KATE WHITEFOOT, Carnegie Mellon University

Staff

CATHERINE WISE, Associate Program Officer
KASIA KORNECKI, Program Officer
JASMINE BRYANT, Program Assistant
ELIZABETH ZEITLER, Associate Director
REBECCA DeBOER, Research Assistant
KYRA HOWE, Research Assistant, *Rapporteur*
RAPHAEL APEANING, Christine Mirzayan Science and Technology Policy Graduate Fellow, *Rapporteur* (until November 2021)

BOARD ON ENERGY AND ENVIRONMENTAL SYSTEMS

[1] Member, National Academy of Engineering.

Reviewers

This Proceedings of a Workshop was reviewed in draft form by individuals chosen for their diverse perspectives and technical expertise. The purpose of this independent review is to provide candid and critical comments that will assist the National Academies of Sciences, Engineering, and Medicine in making each published proceedings as sound as possible and to ensure that it meets the institutional standards for quality, objectivity, evidence, and responsiveness to the charge. The review comments and draft manuscript remain confidential to protect the integrity of the deliberative process.

We thank the following individuals for their review of this proceedings:

Nelson Kelly, Macomb Community College,
Anna Stefanopoulou, University of Michigan, and
Nancy Whiting, Transportation Research Board.

Although the reviewers listed above have provided many constructive comments and suggestions, they were not asked to endorse the content of the proceedings nor did they see the final draft before its release. The review of this proceedings was overseen by John Kassakian, NAE,[1] Massachusetts Institute of Technology. He was responsible for making certain that an independent examination of this summary was carried out in accordance with standards of the National Academies and that all review comments were carefully considered. Responsibility for the final content of this summary rests entirely with the rapporteurs and the National Academies.

[1] Member, National Academy of Engineering.

Contents

Summary of Workshop Presentations

INTRODUCTION

The widespread adoption of electric vehicles (EVs) will play a critical role in decarbonizing the transportation sector as the nation moves toward net-zero emissions. Recent announcements from automakers and the federal government,[1] as well as provisions in the Infrastructure Investment and Jobs Act of 2021,[2] aim to stimulate EV deployment, and ongoing technology improvements continue to make EVs a more affordable and practical option. However, many challenges remain to meet the needs of all buyers and drivers and to ensure that manufacturing supply chains and the electric system can support this large-scale transformation.

As a follow-up activity to its 2021 report *Assessment of Technologies for Improving Light-Duty Vehicle Fuel Economy—2025-2035*,[3] the National Academies of Sciences, Engineering, and Medicine convened a 4-day virtual workshop on October 25-28, 2021, to identify some of the challenges to widespread EV deployment and discuss policy, technical, and market strategies to help federal agencies and other stakeholders plan for the future. The first day of the workshop provided an overview of EV technologies, capabilities, and policy and regulatory considerations—including highlights from related content in the aforementioned National Academies report—and examined the role of EVs within a decarbonized transportation system. The subsequent 3 days included more detailed technical discussions on the following topics:

- *Vehicle production and life cycle,* including impacts to existing supply chains and workforce, new supply chain needs, and end-of-life options such as battery recycling.

[1] (a) B. Preston and J.S. Bartlett, 2021, "Automakers Are Adding Electric Vehicles to Their Lineups. Here's What's Coming," Consumer Reports, November 29, https://www.consumerreports.org/hybrids-evs/why-electric-cars-may-soon-flood-the-us-market-a9006292675. (b) U.S. Department of Energy and U.S. Department of Transportation, 2021, "DOE and DOT Launch Joint Effort to Build Out Nationwide Electric Vehicle Charging Network," energy.gov, December 14, https://www.energy.gov/articles/doe-and-dot-launch-joint-effort-build-out-nationwide-electric-vehicle-charging-network. (c) J.R. Biden, 2021, "Executive Order on Catalyzing Clean Energy Industries and Jobs Through Federal Sustainability," December 8, https://www.whitehouse.gov/briefing-room/presidential-actions/2021/12/08/executive-order-on-catalyzing-clean-energy-industries-and-jobs-through-federal-sustainability.

[2] Atlas Public Policy, 2021, "Infrastructure Investment and Jobs Act (H.R. 3684): Summary of the EV Provisions," Washington, DC: Atlas Public Policy, November 17, https://www.atlasevhub.com/wp-content/uploads/2021/11/2021-11-17_Infrastructure_Investment_and_Jobs_Act.pdf.

[3] National Academies of Sciences, Engineering, and Medicine, 2021, *Assessment of Technologies for Improving Light-Duty Vehicle Fuel Economy—2025-2035*, Washington, DC: The National Academies Press, https://doi.org/10.17226/26092.

- *Challenges and solutions for the electric system,* including the impacts of mass-market EV deployment on electricity generation and distribution systems as well as technological and consumer considerations for EV charging networks.
- *Meeting consumer needs,* including current and projected trends in EV adoption and strategies for making EVs affordable and accessible to all buyers and drivers.

The concluding session on the final day featured presentations on options for removing barriers to EV adoption once EVs have reached cost parity with internal combustion engine (ICE) vehicles.

The following sections provide factual summaries of the workshop presentations. The views contained in this proceedings are those of the individual workshop participants and do not necessarily represent the views of the participants as a whole or of the National Academies. In addition to the summary provided here, materials related to the workshop can be found on the event webpage,[4] including speaker presentations and archived webcasts of the presentations and discussion.

ELECTRIC VEHICLES 101 AND TRANSPORTATION DECARBONIZATION

Steven Cliff, acting administrator of the National Highway Traffic Safety Administration (NHTSA), opened the workshop with keynote introductory remarks. He began by acknowledging the importance of this and past National Academies' activities to NHTSA's efforts promoting road safety and an equitable and environmentally friendly transition to a net-zero greenhouse gas (GHG) emissions transportation sector. He expressed optimism about the maximum feasible targets proposed by NHTSA to improve the Corporate Average Fuel Economy (CAFE) Standards for Model Years 2024-2026 Passenger Cars and Light Trucks. These standards would save billions of dollars for consumers and provide crucial environmental benefits by reducing reliance on fossil fuels. Cliff then turned to the subject of vehicle electrification, describing NHTSA's research and policy focus on resiliency and cybersecurity issues of vehicle battery management systems.

Gary Marchant, the Regents' Professor of Law and director of the Center for Law, Science, and Innovation at Arizona State University, provided an overview of the EV landscape within the context of the workshop's themes. He explained that the workshop was prompted by the 2021 National Academies study on light-duty vehicles (LDVs),[5] which provides a broad consensus on the goal, urgency, and ability for a major transition to zero-emission vehicles (ZEVs) by 2035. The transition to a U.S. LDV fleet dominated by ZEVs would involve an unprecedented system-wide change that will affect every aspect of the transportation sector. Many critical and complex issues need to be addressed to manage this transition, including battery capabilities, infrastructure needs, supply chain requirements, regulatory alignment, and more. Marchant then noted that the accelerating momentum toward ZEVs in the United States is primarily due to improvements in battery price and capabilities. In addition, while many U.S. vehicle manufacturers have committed to plans for a transition to an all-electric fleet, recent trends indicate the United States is lagging behind China and Europe in the deployment of EVs (see Figure 1). Marchant specified that there is an impetus for the United States to catch up, not only from an environmental perspective, but also for competitive reasons. He then turned to an important finding from the aforementioned National Academies' report pointing to consumer unfamiliarity and misperception as a significant barrier to the adoption of battery electric vehicles (BEVs). Congress could help overcome these barriers by providing consumer incentives and the electrification infrastructure; further, Congress's intervention will be required to help resolve technical and legal issues related to ZEV fuel economy. Marchant concluded that despite the challenges, the EV transition is worth pursuing as a society.

Nady Boules, president of NB Motors, LLC, began his presentation by outlining the current state-of-the-art BEV drive systems. After explaining the various components of the EV propulsion drive, Boules emphasized that

[4] National Academies of Sciences, Engineering, and Medicine, 2021, "Assessment of Technologies for Improving Fuel Economy of Light-Duty Vehicles—Phase 3 Electric Vehicles Workshop," October 25-28, https://www.nationalacademies.org/event/10-25-2021/assessment-of-technologies-for-improving-fuel-economy-of-light-duty-vehiclesphase-3-electric-vehicles-workshop.

[5] National Academies of Sciences, Engineering, and Medicine, 2021, *Assessment of Technologies for Improving Light-Duty Vehicle Fuel Economy—2025-2035*, Washington, DC: The National Academies Press, https://doi.org/10.17226/26092.

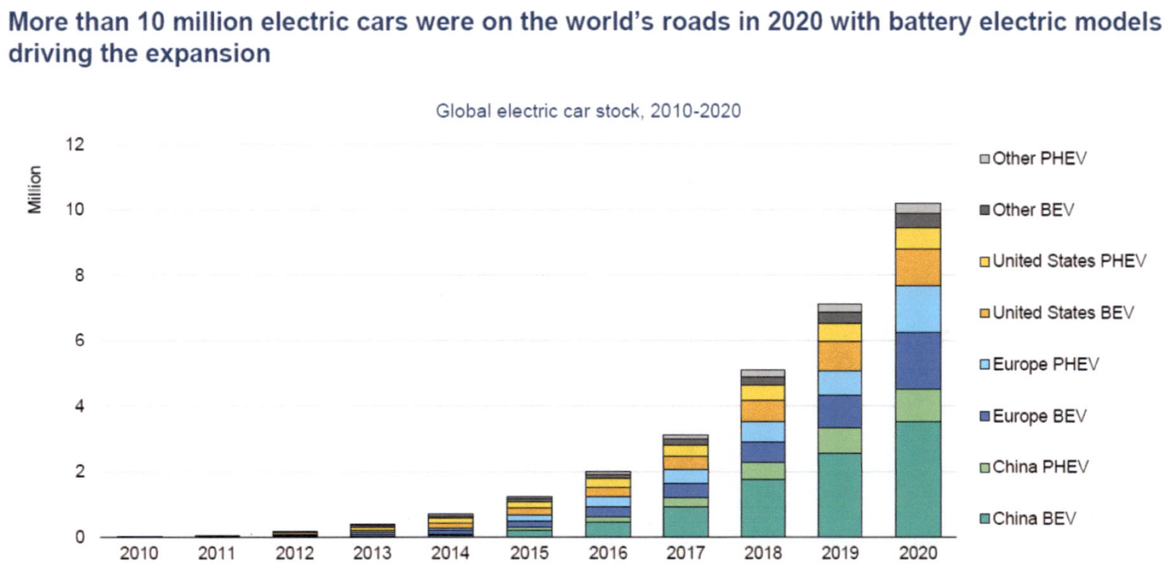

FIGURE 1 Total electric vehicle (EV) registrations and sales in China (green), the United States (yellow), Europe (blue), and other major markets. NOTES: While EV registrations and sales in most major markets increased in 2020, EV deployment in the United States continues to lag behind China and Europe. The selected countries are the largest EV markets and are ordered by size of the total car market in the upper half of the figure and by sales share of electric cars in the lower half. Regional EV registration data can be interactively explored via the Global EV Data Explorer. BEV, battery electric vehicle; PHEV, plug-in hybrid vehicle. SOURCES: Gary Marchant, Arizona State University, "Overview of the Electric Vehicle Landscape," presentation to the workshop, October 25, 2021; from International Energy Agency, 2021, "Global EV Outlook 2021: Accelerating Ambitions Despite the Pandemic."

these components are mature technologies and have been refined over decades. The application challenges of these components lie in the need to meet the high torque and power density requirements of EVs. He noted the EV industry has converged on using the permanent magnet synchronous motor drive with rare earth magnets due to its superior efficiency, torque, and power density, despite its relatively higher cost. Boules discussed various strategies to improve performance and reduce the cost of permanent magnet synchronous motors, such as increasing motor speed and using less rare earth magnet material. Boules concluded by providing projected performance improvements associated with advanced motors and power electronics.

Deidre Strand, chief scientific officer at Wildcat Discovery Technologies, discussed the fundamentals of lithium-ion (Li-ion) batteries and explained what is required to improve their performance for EV applications. She described how lithium ions provide a balance between energy and power, making lithium the most suitable element for EV battery applications. Cost, safety, and lifetime are also important performance metrics for automotive batteries. Moreover, an improvement in the performance of a single parameter is often at the detriment of other performance parameters of the battery. For instance, replacing the battery's active material to improve power, energy density, and lifetime have associated tradeoffs in cost, stability, and supply chain risk that must be managed. Strand emphasized that although the current battery technologies are safe, safety should be an imperative criterion, especially in cases of improving the energy density of batteries. As depicted in Figure 2A, material costs—including those from the cathode, anode, and electrolyte—represent a significant portion of the total cost of the Li-ion cell. Using lower-cost materials, thicker and denser electrodes, less moisture sensitive material, and eliminating or minimizing electrode formation time can be used to improve the performance of Li-ion batteries. In closing, Strand remarked that Li-ion batteries will continue to be the dominant battery technology for EVs and engineering improvements at the module- and pack-level will contribute substantially to EV performance and cost. Battery pack cost projections for 2018-2030 from several studies and automaker statements are shown in Figure 2B.

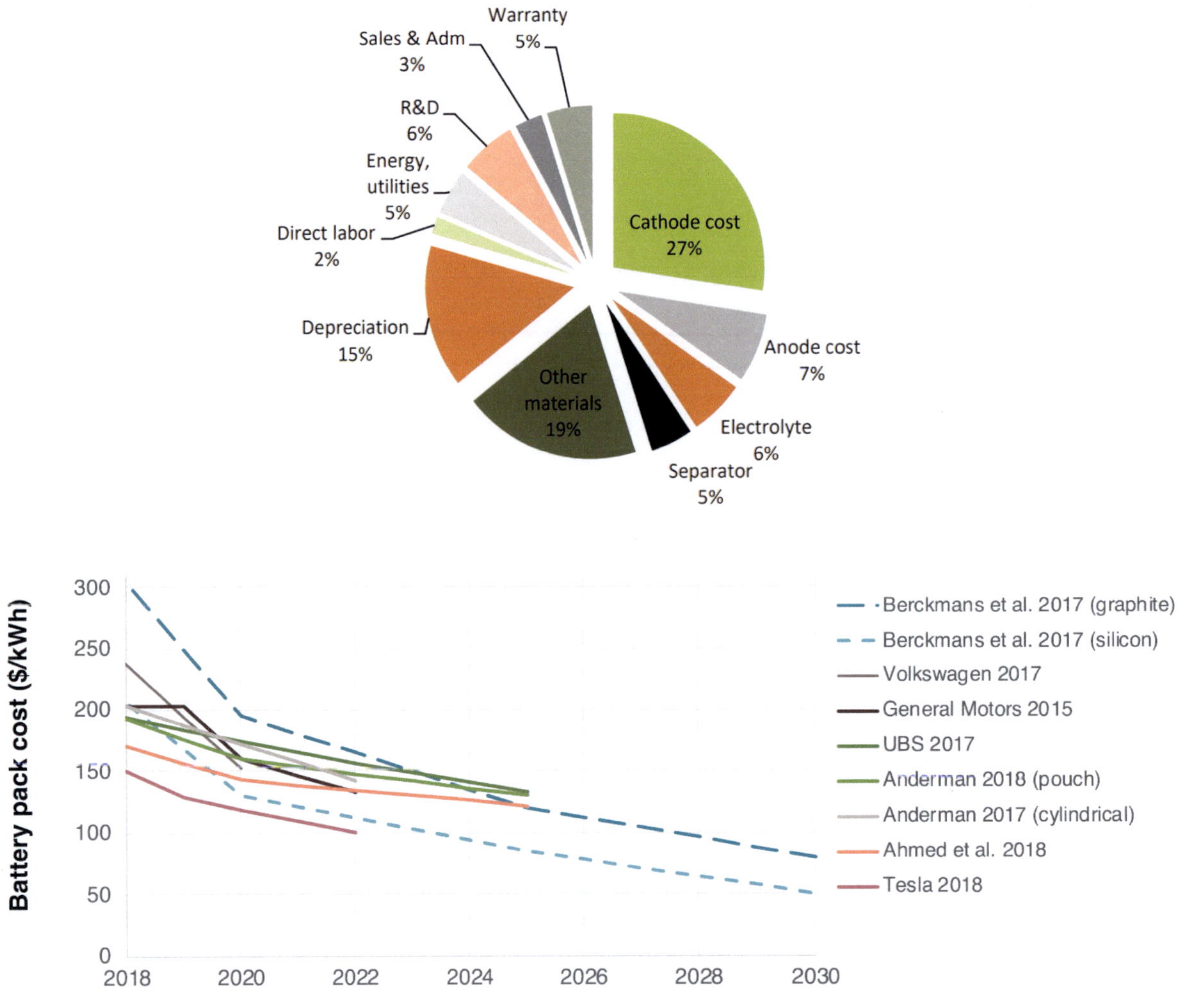

FIGURE 2 (A) Breakdown of costs for lithium ion battery cells, showing that the materials costs—including the cathode, anode, and electrolyte costs—are a significant portion of the total cost structure. (B) Cost projections for electric vehicle battery packs from several technical studies and automaker statements, showing estimated decreases in pack costs over time. SOURCES: Deidre Strand, Wildcat Discovery Technologies, "Introduction to Lithium Ion Batteries," presentation to the workshop, October 25, 2021; from (A) C. Pillot, 2019, "The Rechargeable Battery Market and Main Trends 2011-2020," presented at Avicenne Energy, Paris, France, May 28, http://cdn.ceo.ca.s3.amazonaws.com/1em2t4r-02%20-%20Presentation%20 Avicenne%20-%20Christophe%20Pillot%20-%2028%20Mai%202019.pdf; (B) Modified from N. Lutsey and M. Nicholas, 2019, "Update on Electric Vehicle Costs in the United States Through 2030," International Council on Clean Transportation, licensed under CC BY-SA 4.0, https://creativecommons.org/licenses/by-sa/4.0.

A presentation by **Denise Gray**, president of LG Energy Solution Michigan Inc. Tech Center, provided a general overview of the business ecosystem of battery manufacturing. She began by citing ongoing battery manufacturing collaborations between automotive and electronic companies. The business models for these collaborations are often direct purchase or joint venture agreements. Gray added that there are ongoing efforts to develop new business models to secure the supply of battery technologies to meet the growing demand for EVs. These business models consider not only the needs of the manufacturer but also the business needs of partners along the battery technology supply chain. Satisfying the business and operation needs of other sub-suppliers and partners is crucial to sustaining and securing the supply of batteries. Gray spoke about the importance of stakeholders at the state, federal, and local levels in helping to secure the investment infrastructure required to deploy battery manufacturing facilities. In addition, she acknowledged the role laboratories, universities, and automotive manufacturing firms have played to improve the cost-effectiveness of battery technology. Moving forward, the future cost and efficiency improvements in battery technologies cannot be achieved in isolation but will require system integration with other components of the EV.

A presentation by **AK Srouji**, chief technology officer at Romeo Power, complemented those of Strand and Boules by providing a state-of-the-industry perspective on EV battery and electronic management systems. He discussed the battery system in terms of four main pillars: Cell Science, Module Technology, Pack Technology, and Battery Management Systems (BMSs). From the cell science point of view, a resurgence of lithium iron phosphate (LFP) is becoming apparent for vehicles with low to medium range, while nickel-rich cells (nickel-cobalt-manganese and nickel-cobalt-aluminum) remain the choice for longer-range vehicles, necessitating highest energy density. Thermal management begins with module technology that works to maintain the battery temperature within the ideal operating window of 25 to 35 degrees Celsius, which is crucial for ensuring the battery lasts the lifetime of the vehicle. Additionally, pack technology incorporates safety features and mitigates risk from crashes and other large loads, such as vibration, and ensures protection from the environment (water intrusion, humidity, debris). Lastly, the BMS functions as the brain of the battery by overseeing control functions, managing safety-critical decisions, and evaluating battery states, such as state of charge (SOC) and state of health (SOH). In closing, Srouji discussed some key performance indicators for battery system energy density and safety, emphasizing the importance of safety features such as single cell fault tolerance, whereby thermal runaway in a given cell stays contained and does not spread to other neighboring cells.

Dave Cooke, senior vehicles analyst at the Union of Concerned Scientists, discussed the need to rethink transportation decarbonization through the lens of local communities most affected by the impacts of fossil fuel use and climate damage; this implies a reconsideration of the policies and interventions that are incentivizing the adoption of EVs. Cooke emphasized that new car buyers are generally not representative of communities being affected by the adverse impacts of fossil power. He further illustrated how eligibility caps have, in the past, incentivized incremental sales while excluding the neediest demographic for whom new vehicle purchase is not a preferred option. Cooke cited the fact that California's priorities for broader clean vehicle adoption have moved toward a more targeted program that better addresses these inequities. Another critical policy challenge is inequalities associated with federal support for EV charging. Since home charging remains the dominant mode of charging EVs, households living in multi-unit dwellings are at a significant disadvantage to access charging. Even with public-funded charging, there is evidence from California that these sites tend to track the demographic of the early EV adopter market rather than those of the communities most affected by car and truck pollution. Moving forward, new policy paradigms are needed that provide solutions for communities that stand to benefit the most from the adoption of EVs, tackling long-standing inequities and justice while reckoning with climate change. Cooke concluded with a discussion of policies for EVs and electric charging that could achieve that goal, including new vehicle regulations, accessible financing, used-vehicle incentives, early vehicle retirement programs, filling charging information gaps, providing charging for communities not passersby, and addressing building codes.

A presentation by **Alexander Edwards**, president of Strategic Vision, provided insights into the consumer psychology impacts associated with BEV adoption. He began by overviewing the significant barriers to owning a BEV (see Figure 3). For most new car buyers in the United States, the ancillary cost and compromise associated with BEVs often discourages their purchase. In contrast, among early EV adopters, the extra cost and compromises of EVs are often overlooked in favor of advantages such as quiet performance, powerful torque, and near-instant acceleration. Edwards illustrated how changes in personal mobility needs and working conditions due to the COVID-19 pandemic have increased consumer interest in BEVs. With this interest, the EV industry is presented with an opportunity to capture the hearts and minds of potential consumers. Edwards remarked that different subgroups within the BEV consumer market have varying EV feature preferences; for example, he noted that most current BEV buyers purchase the vehicles for their superior performance and acceleration compared to ICE vehicles, but other individuals purchase BEVs for environmental reasons or because they want to own an "innovative, smart, and cool" vehicle. Edwards concluded on the note that BEVs are in a very competitive landscape with ICE vehicles and all other alternative powertrain vehicles.

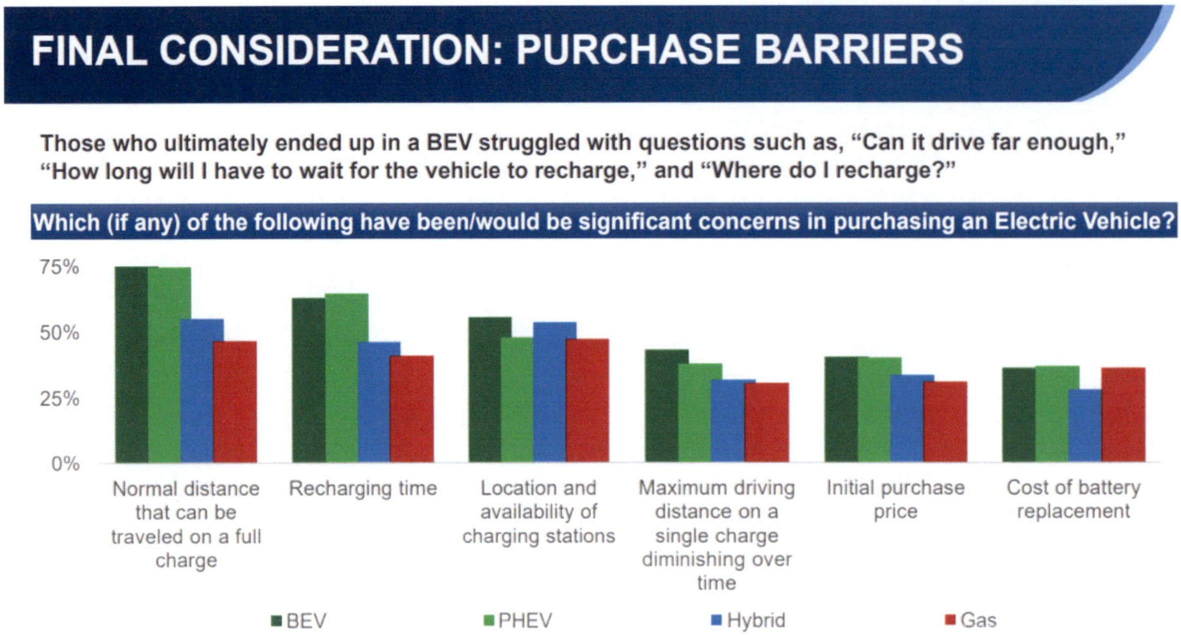

FIGURE 3 New Vehicle Experience Study survey responses from battery electric vehicles (BEVs), plug-in hybrid electric vehicles (PHEVs), hybrid and gas vehicle owners, and their perceptions of the major electric vehicle purchase barriers. SOURCE: Alexander Edwards, Strategic Vision, "Alternative Energy and Mobility," presentation to the workshop, October 25, 2021.

Kate Whitefoot, associate professor of mechanical engineering and engineering and public policy at Carnegie Mellon University, began her presentation by highlighting the need to evaluate the potential behavior of mainstream BEV adopters considering the fact that future BEV consumers will be different from current early adopters. Representing a major shift in technology and consumer behavior, the adoption of BEVs may be impeded by numerous barriers, including limited range, risk aversion, and unfamiliarity with BEV technology. Improving the range of EVs as well as increasing consumer familiarity can help overcome these barriers. In her concluding remarks, Whitefoot asserted that mainstream adoption of BEVs is expected as they near cost parity with their equivalent gasoline counterparts. To reflect this increasing market share, federal agencies should begin to consider ZEVs when setting vehicle efficiency and emissions standards. In response to a public question about consumer "range anxiety," Whitefoot emphasized the need for more research looking at multivehicle households, and suggested consumers may be more comfortable using their BEVs for short range commuting and their ICE vehicle for longer distance trips.

Joshua Linn, associate professor at the University of Maryland and senior fellow at Resources for the Future, focused on EV consumer adoption behavior and the effectiveness of policy instruments, such as subsidies in promoting the adoption of EVs. Citing a study by Leard et al. (2019), Linn explained how federal and state subsidies account for about half of new EV sales in 2019. Using the same reference, he explained how lower-income groups are more responsive to the prices of EV compared to high-income groups. Linn then discussed the effectiveness and impacts of three possible scenarios for extending EV subsidies, indicating that low-income subsidies tend to be more cost-effective in driving EV purchases, given that low-income consumers tend to have a lower EV demand and are more price sensitive. Linn concluded by discussing the interplay and interaction between subsidies and other emission and efficiency policy targets, asserting that income-based subsidies would be regressive in the short run because of interactions with other policies.

Rachael Nealer, deputy director for transportation technology and policy for the White House Council on Environmental Quality, outlined a selection of federal investments and policy interventions targeting transportation decarbonization. She emphasized the Biden administration's commitment to tackling the climate crisis by decarbonizing all sectors across the economy to reach the overall goal of net-zero emissions by 2050. Nealer then highlighted the urgency of scaling up EV adoption to meet both near- and long-term transportation decarbonization goals. The Biden administration will make investments in response to the Infrastructure Investment and Jobs Act to ramp up transportation electrification in alignment with the administration's Justice40 Initiative on environmental justice. The same bill has also allotted provisions for grants that support electrification, equitable deployment of EV network infrastructure, and electrification of school bus fleets. Additionally, the federal government has a role in regulating the policy landscape of transportation sector decarbonization. Aviation decarbonization has been a priority for the Biden administration and in the near-term, meeting mitigation targets will require partnerships that spur the deployment of billions of gallons of sustainable aviation fuel within the decade. In closing, Nealer articulated that decarbonization of the transportation sector is required in order to address the climate crisis.

A presentation by **Patty Monahan**, commissioner at the California Energy Commission (CEC), provided an overview and outlook of California's climate change policies with a focus on zero-emission transportation. She began with a brief historical background on California, highlighting the state as a forerunner in the adoption of zero-emission transportation policies. She followed the historical context with a discussion on California's clean electricity goals, noting that meeting the goal of 100 percent zero-carbon electricity by 2045 would also benefit transportation decarbonization, since there would be no emissions associated with BEV charging. Furthermore, Monahan illustrated how the burden of transportation-related pollution is disproportionately borne by low-income and minority communities (see Figure 4), and stressed the need to address these disproportionate impacts not only in California, but across the United States. She then discussed California's EV goals and budget priorities targeting these goals. Despite the upward trajectory of California's ZEV adoption since 2011, there is a massive scale-up

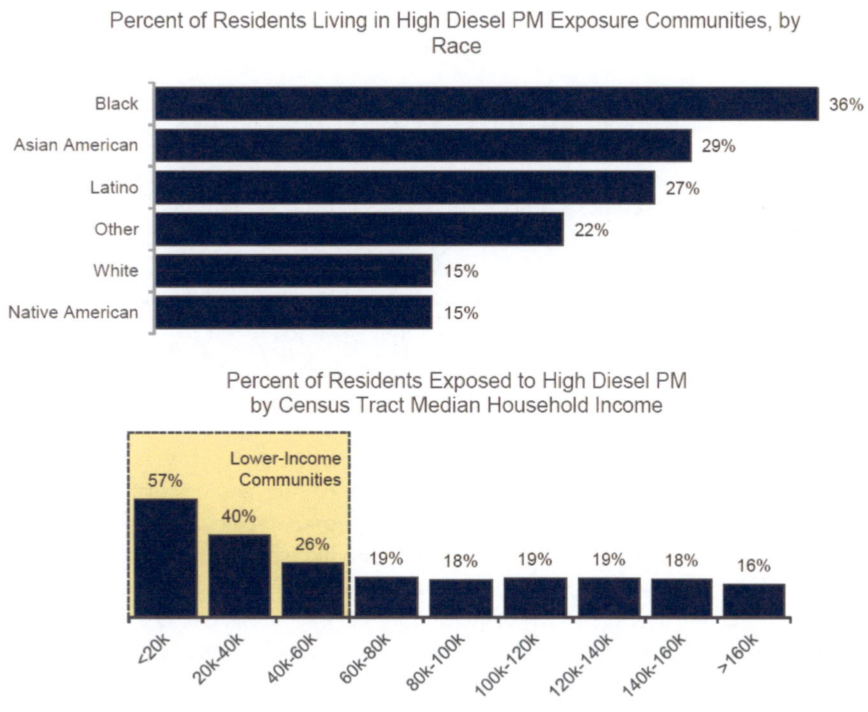

Credit: CEC analysis of census and CalEnviroScreen data (Note: "High Diesel PM Exposure" communities are census tracts that score in the highest 75th percentile of census tracts for diesel particulate matter. The vast majority [90 percent] of diesel PM emissions come from vehicles.)

FIGURE 4 Plot of the share of California residents exposed to high levels of diesel particulate matter by race and household income. The most affected populations tend to be Black, Asian American, Latino, and lower-income residents. SOURCES: Patty Monahan, California Energy Commission, "California: Leading the Future of Transportation," presentation to the workshop, October 25, 2021; from Q. Gee, S. Bailey, J. Berner, M. Comiter, J. McKinney, and T. Olson, 2021, *Final 2020 Integrated Energy Policy Report Update*, Publication Number CEC-100-2020-001-V1-CMF, California Energy Commission, Sacramento, Calif.

gap in passenger vehicle charging infrastructure that needs to be reached by 2030. As the plug-in electric vehicle (PEV) market grows, investing in school buses with vehicle-to-grid capability and managing charging impacts on the grid to maximize renewable energy use will be critical.

Daniel Sperling, director of the Institute of Transportation Studies at the University of California (UC), Davis, discussed the trajectory and potential impact of EVs from a policy, regulatory, and academic perspective. He began by emphasizing that the future of EVs is mostly socially determined, in that the market penetration of EVs will be dependent on policy and consumer choice. As such, the deployment of EVs is a policy imperative for decarbonizing the U.S. economy. For example, aggressive low-carbon surface transport policies will result in carbon emissions reductions, cost savings, and lower health costs. Sperling reiterated the insights from the previous speaker about how consumers do not make their purchase decisions based solely on the total cost of ownership; rather individual buyers are conservative about their choices and are concerned about factors such as resale value, the future cost of energy, loss aversion, and range anxiety. Given these challenges, incentives will be needed to spur the adoption of EVs. These incentives do not necessarily have to be borne by the taxpayer; instead, innovative schemes such as feebates[6] could be considered. Using insights from consumer behavior research, Sperling explained that capturing the first 50 percent of LDV market share will be relatively easy compared to the last 30 percent due to consumer heterogeneity in ideology and preferences. Government assistance and funding will be required to ensure the charging infrastructure is accessible and reliable to many consumers, and to spur innovation and transform the business model around charging infrastructure. Sperling added that most legacy U.S. automakers are holding back on the production of EVs; thus, aggressive regulations will be an important lever necessary to accelerate EV deployment. In his closing remarks, Sperling emphasized the need for EV policy incentives to promote vehicle purchases and send a clear market signal to consumers.

VEHICLE PRODUCTION AND LIFE CYCLE

Kristin Dziczek, senior vice president of research at the Center for Automotive Research, began her presentation with a discussion of the vulnerabilities and sensitivities of the U.S. automobile industry to supply chain disruptions stemming from factors such as natural disasters, labor disputes, shipping disruptions, and more. For example, Dziczek warned that increasing BEV/PHEV production while decreasing ICE production could lead to low productivity and potential plant shutdowns and consolidations. She suggested that suppliers tied to ICE engines will make fewer and fewer parts as ICE LDV production declines, while the scale of manufacturing EV parts and components will remain insufficient for some time. Dziczek noted that the novelty of EV production lines is disrupting the entire ecosystem of the automobile industry. This includes upending existing supply chains with new entrants in the form of joint ventures, mergers and acquisitions, and new leaders. Automakers are adopting new strategies to cater to these supply chain and material sourcing disruptions caused by the EV transition. Dzieczek added that the application of multi-use platforms, propulsion components, and uniform design architecture are a few examples of strategies employed by automakers to reduce EV manufacturing complexity. In her concluding remarks, Dzieczek suggested that as PEVs disrupt the automobile industry, both supplier and manufacturer will have to adapt to changes in the supply-chain needs of the BEV industry. When asked about semiconductor supply chain issues, she remarked that the average [ICE] vehicle has $300 in semiconductor content compared to nearly $3,000 in some BEVs. Building new capacity at the scale needed will take time and costs will be steep, at ~$12 billion to $14 billion for a new semiconductor plant.

Brad Markell, executive director of the AFL-CIO Industrial Union Council, opened by emphasizing that the switch to EVs might have some very deep and potentially troubling implications for communities and the labor economy at large. He alluded to the fact that the barriers to entry in the EV industry are relatively low compared to the traditional automotive industry due to the EV market's lower ratio of capital to labor requirement. Markell emphasized that the switch to EVs might create a lot of opportunities for automation and simpler assembly lines.

[6] A feebate for energy efficient products incentivizes purchase of efficient products and disincentivizes purchases of inefficient products. The purchasers of efficient products are provided a rebate, and purchasers of inefficient products are charged a fee. The revenues from fees fund rebates.

These opportunities for new players could potentially disrupt the automobile labor economy. Markell closed his presentation by remarking that the successful ramping up of EVs will require the full suite of government support and policies that promote the rapid transitions to EVs while ensuring the domestic inflow of the labor benefits.

Zoe Lipman, director of manufacturing and advanced transportation at the BlueGreen Alliance, discussed the policy actions required to strengthen automobile manufacturing and supply chains. Lipman emphasized that pro-climate policies are aligned with pro-worker and environmental justice agendas. Advanced manufacturing and supply chain infrastructure for the EV industry will play an important role in the transition to a clean fuel economy. The auto industry is at a crossroads, with both domestic and international players moving swiftly to shape the competitive landscape. Sound policies will be required to shape the competitive edge of the U.S. economy in capturing the manufacturing and supply chains of EVs. Lipman believes there are tremendous opportunities to affect the outcomes of jobs through policies that ensure beneficial gain to the U.S. economy. She further stated that about 220,000 manufacturing jobs hinge on whether the shift to EVs takes place, together with policies that ensure the on-shoring of the EV supply chain and production. Lipman concluded that the United States can ensure a successful EV transition by investing in manufacturing retooling and conversion, re-shoring, and filling critical supply chain gaps.

Warren Day, research geologist with the U.S. Geological Survey (USGS) Mineral Resources Program, provided an overview of the critical minerals and resources needed for EV manufacturing. He noted that the United States is highly import-reliant on a large and growing number of mineral commodities required for EV manufacturing.[7] The USGS is undertaking several initiatives that provide access to basic geoscience information, which can help accelerate a better understanding of the mineral endowment of the United States. Day provided the example of the ongoing Earth Mapping Resources Initiative (Earth MRI), which seeks to improve our knowledge of the U.S. geologic framework in areas permissive for hosting undiscovered critical mineral resources. Figure 5, for example, shows the locations of significant lithium deposits. This generation of geoscience data is the first step in a 5-10-year exploration and licensing process that culminates in the opening of a mine. Additionally, Day discussed the details of GeoDaWN, a geoscience data acquisition project in Western Nevada aimed at collecting high-resolution airborne magnetic, radiometric, and light detection and ranging (LIDAR) data to understand the region's geology, natural resources, and geologic hazards. Day concluded his presentation by indicating that the USGS aims to collect geoscience data on subsurface mineral resources from unconventional sources, such as mine waste, to recover additional critical mineral commodities.

David Klanecky, executive vice president and chief operating officer at Piedmont Lithium, elaborated on the importance and opportunities of localizing the lithium-ion battery supply chain within the United States. Klanecky began by discussing the market dynamics of lithium, pointing to the fact that demand is expected to grow significantly in the future with high EV penetration as the United States moves toward electrifying its transportation systems. He highlighted the projected domestic shortfalls in the supply of lithium to meet this increased demand for lithium-ion battery technologies and a call to action by companies and the U.S. government to address this challenge. Additionally, he noted that about 90 percent of the global supply of lithium hydroxide material used in high energy density lithium-ion batteries is being produced in China. Klanecky explained that there is growing interest in the United States and Europe for the localization and regionalization of the lithium supply chain for the reasons of supply security, lowering cost, increasing speed to market, and reducing environmental footprints. If the supply chain of lithium batteries is not localized, the United States stands to lose about $100 billion by 2040 due to the need of imported materials. Klanecky concluded by emphasizing the importance of localizing the lithium supply chain from a national and energy security perspective.

Celina Mikolajczak, vice president of manufacturing engineering at QuantumScape, provided a general overview of the supply chain needs for lithium battery production. The typical lithium-ion cell supply chain is a sophisticated ecosystem made up of specialized factories that create the highly refined and engineered components of batteries. Mikolajczak noted that the factory equipment supporting lithium batteries is also complicated, as it operates under conditions of high production rates and holds micron-level tolerances under

[7] See U.S. Geological Survey, 2021, "Mineral Commodity Summaries 2021," https://doi.org/10.3133/mcs2021, which lists manganese, cobalt, graphite, and rare earth elements as among the critical minerals needs.

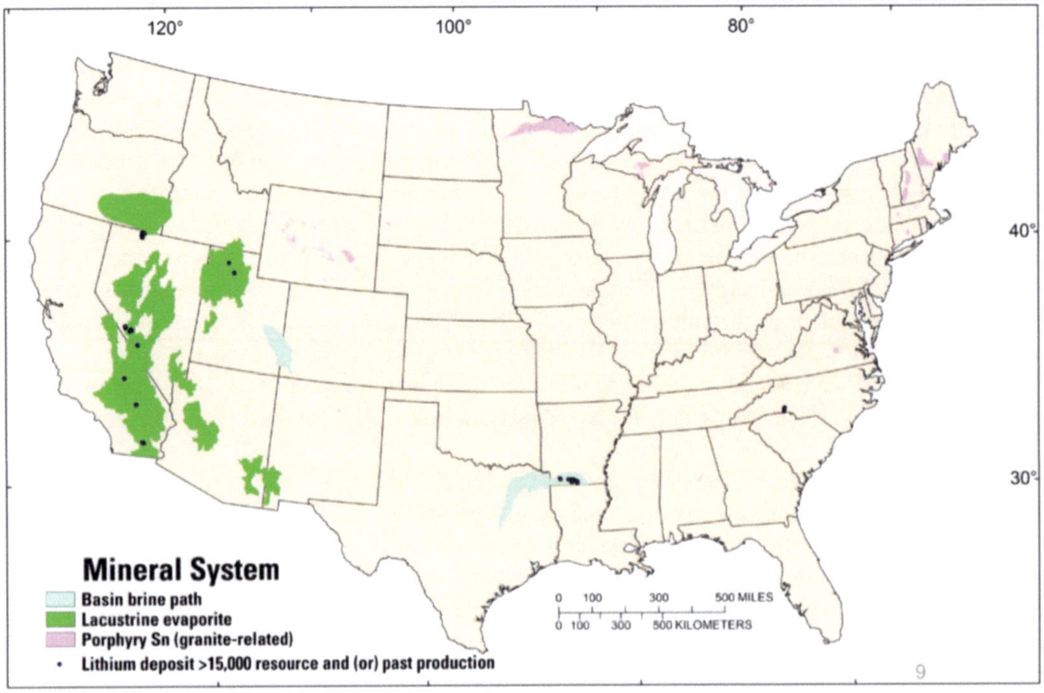

FIGURE 5 Locations of basin brine (blue), lacustrine evaporate (green), porphyry tin (pink), and lithium deposits (black) in the contiguous United States. The geological conditions of these systems are ideal for lithium formation. SOURCE: Warren Day, U.S. Geological Survey, "Overview of the USGS Earth Mapping Resources Initiative Geoscience Data for the Upstream Critical Minerals Supply Chain," presentation to the workshop, October 26, 2021.

extreme cleanliness requirements. Moreover, the operation and maintenance of cell-making equipment and factories require highly skilled machinists, mechanics, and engineers. Mikolajczak concluded with a discussion of two major challenges facing this industry: the near total absence of U.S. suppliers of engineered cell materials and cell-making factory equipment; and the lack of sufficient production and engineering staff in cell-making factories. During the question and answer session, she elaborated that factories supporting the lithium battery supply chain will likely be sited near resources, the ultimate customer, or sites with abundant energy available for the factory processes.

Linda Gaines, chief scientist of the ReCell Center for Battery Recycling at Argonne National Laboratory, (ANL) began by introducing ReCell, a U.S. Department of Energy (DOE) initiative seeking to significantly reduce U.S. dependence on critical raw material imports by use of recycled material feedstocks. Gaines emphasized Mikolajczak's point that critical deficiencies exist in our ability to purify and refine materials along the U.S. lithium battery supply chain. Limited reserves of critical resources and the lack of U.S. processing capacity highlight the importance of recycling. Even with recycling as a viable material source, the United States would have to develop the manufacturing infrastructure to handle intermediate products. Gaines defined the direct recycling of batteries' critical material as a process of recovering and reusing cathode material without breaking down its chemical structure. Direct recycling represents the lowest impact and highest revenue option compared to pyrometallurgical and hydrometallurgical recycling processes (see Figure 6). Gaines noted that scrap from battery cell manufacturing is currently the most viable feedstock for direct recycling because it does not require purification and is less energy-intensive to process. In addition to the economic gain, the direct recycling of cathodes can reduce significant environmental impacts in various categories such as GHGs and water use. Gaines concluded her presentation by highlighting some critical challenges and barriers facing the recycling industry, including, but not limited to, securing industry buy-in for commercialization and developing novel recycling processes for future battery technologies.

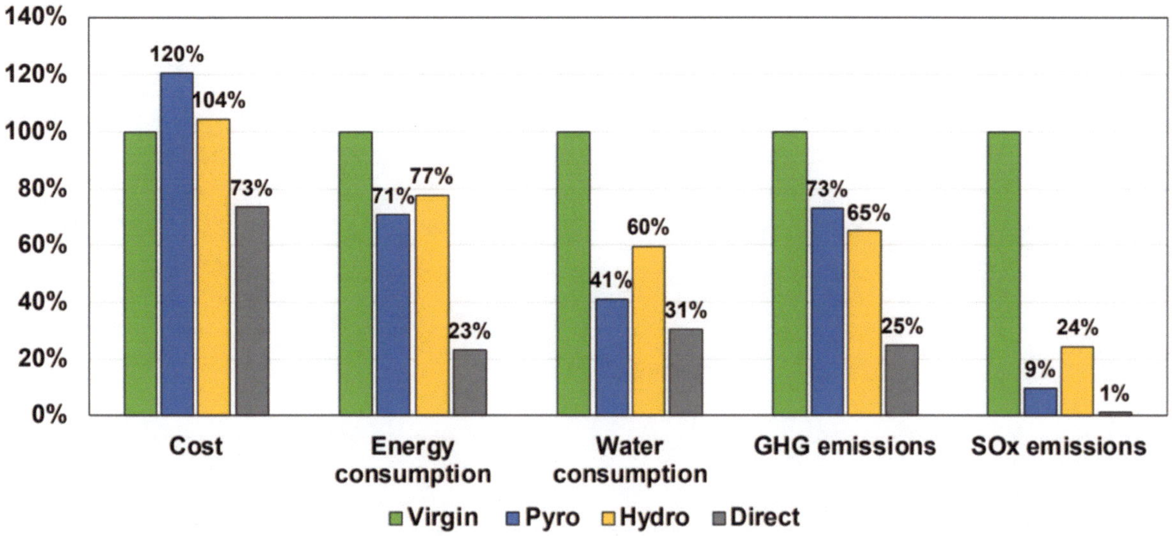

FIGURE 6 Comparison of cost and environmental impacts associated with the production of 1 kg of lithium nickel manganese cobalt oxides (NMC111). These processes include the baseline extraction of virgin ore (green), pyrometallurgical process recycling (blue), hydrometallurgical process recycling (yellow), and direct recycling (grey). SOURCE: Linda Gaines, ReCell Center for Battery Recycling, "Li-ion Battery Recycling and the ReCell Center," presentation to the workshop, October 26, 2021.

A presentation by **Fan Dai**, executive director of the California-China Climate Institute at UC Berkeley highlighted the current and future EV battery reuse and recycling policies in China. Dai began with an overview of the EV battery industry in China, noting that battery reuse and recycling policies will play a crucial role as China reaches higher volumes of ZEV sales. She also indicated that there is an increasing shift from nickel-manganese-cobalt-oxide batteries to LFP batteries due to safety, lower cost, consumer preference for shorter-range vehicles, and domestic manufacturing capabilities. Dai then discussed two policy initiatives in China that aim to identify the requirements of companies that use battery packs from retired EVs and encourage cooperation between battery manufacturers and reuse companies. Dai concluded her presentation by indicating that the United States is lagging behind China and the European Union in establishing policy guidelines for EV battery reuse and recycling at the national level; to this end, drawing lessons from China and the European Union could help to establish a more sustainable EV supply chain in the United States. As a follow up to a public question about battery passports and state of health information, Dai stressed the need for original equipment manufacturers (OEMs) to capture the right business model in terms of repurposing, reusing, and recycling used batteries, also noting the fact that around 15GW of batteries will need to be recycled in China in 2030.

Building on the previous presentations, **Anna Stefanopoulou**, the William Clay Ford Professor of Technology at the University of Michigan, discussed the opportunities and challenges of battery middle-life from the perspectives of cost of degradation and the value of prognostics. Stefanopoulou noted that there are good reasons for paying attention to the middle-life of lithium-ion batteries as it pertains to decisions about repair, reuse, refurbishment, and repurposing. Moreover, she alluded to the fact that this is a large decision problem, and making such a decision about middle-life processes will require a significant amount of information gathering and field data collection to understand the monitoring and management for recycling and repurposing. Manufacturers have acted accordingly by developing telematics to collect field data; however, that data are irregular and the sensors tend to introduce noise biases among other engineering issues. Moving forward, the ability to project and extrapolate the performance of cells, packs, and modules will be critical to making a business decision for reuse and repair (see Figure 7). The projection of capacity degradation is particularly important in this regard, both for cost replacement warranties to the user and companies, and for the environment. Stefanopoulou added that even in a laboratory, assessing the

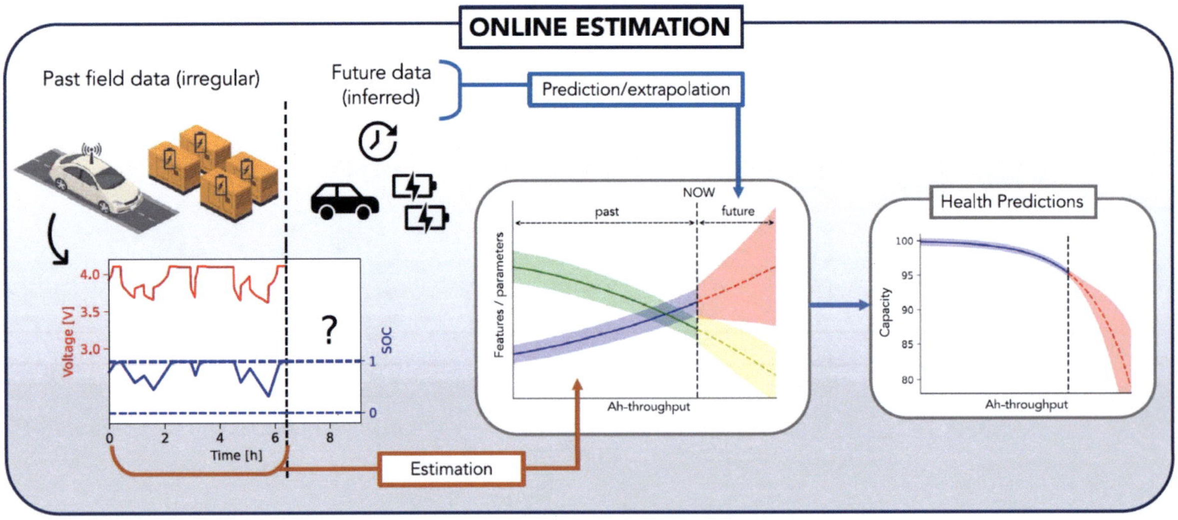

FIGURE 7 Diagram of the steps needed to make the business case for whether to recycle, repair, reuse, refurbish, or repurpose lithium-ion batteries. These steps include collecting field data on battery performance and health, and extrapolating that data to make predictions about how cells age. SOURCES: Anna Stefanopoulou, University of Michigan, "Opportunities and Needs in Battery Middle-Life," presentation to the workshop, October 26, 2021; reprinted from Joule, 5/8, V. Sulzer, P. Mohtat, A. Aitio, S. Lee, Y.T. Yeh, F. Steinbacher, M. Umer Khan, J. Woo Lee, J.B. Siegel, A.G. Stefanopoulou, and D.A. Howey, The challenge and opportunity of battery lifetime prediction from field data, 1934-1955, Copyright 2021, with permission from Elsevier.

degradation modes for groups of cells has proven difficult. She believes developing a stronger understanding of how to adapt and control system degradation will ultimately result in more opportunities to stretch the utilization of the batteries. In closing, Stefanopoulou highlighted the importance of monitoring the SOC and the SOH in the context of performance and the cost of materials.

David Howell, acting director of the Vehicle Technologies Office (VTO) in the Office of Energy Efficiency and Renewable Energy at DOE, opened by outlining the strategies of VTO aimed at decarbonizing transportation across all modes. He discussed President Biden's Executive Order (EO) on America's Supply Chains (EO 14017),[8] focusing on the high-capacity battery supply chain in particular, and highlighted key vulnerabilities in U.S. lithium-ion battery supply chains, including the supply of class 1 nickel, lithium, and cobalt; the major upstream deficit in mineral refinement and processing; cathode and anode production in the mid-stream; and the present lag in domestic demand and recycling for lithium-ion batteries downstream of the supply chain (see Figure 8). Howell outlined key recommendations associated with high-capacity battery supply chain reports, including stimulating demand, strengthening key battery mineral supplies, promoting battery materials, cell and pack production, and investing in people and innovations. Howell concluded by highlighting key innovation focus areas of the VTO to decrease cost, reliance on critical minerals, and charging time for EV battery cells.

Renata Arsenault, technical expert for advanced battery recycling at Ford Motor Company, began by providing key projections that underscore the expected increase in the demand for battery cells and materials. She pointed to the fact that China has gained a strategic advantage by capturing dominance across all processes along the battery supply chain, which poses a risk to U.S. energy security (see Figure 9). Arsenault suggested that there has been disproportionate investment in downstream market segments at the expense of upstream infrastructure in the United States. Mobilizing the battery supply chain will require the deployment of novel supply chains and additional investment in upstream and midstream infrastructure. Arsenault noted that while the recycling of scrap material from aged EVs can play a strategic role in offsetting the primary demand for battery raw materials, there

[8] J.R. Biden, 2021, "Executive Order on America's Supply Chains," February 24, https://www.whitehouse.gov/briefing-room/presidential-actions/2021/02/24/executive-order-on-americas-supply-chains.

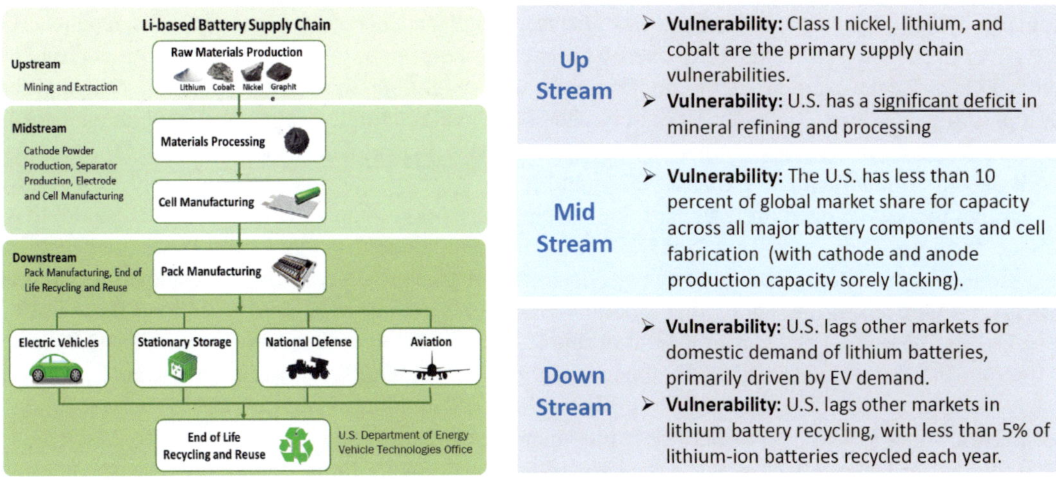

FIGURE 8 Schematic of the lithium-based battery supply chain and major U.S. vulnerabilities in the upstream, midstream, and downstream segments. SOURCE: David Howell, Vehicles Technology Office, "National Blueprint for Lithium Batteries and Executive Order High-Capacity Battery Supply Chain Report," presentation to the workshop, October 26, 2021.

is a noteworthy lag in the timing and volume for the supply of recycling feedstock. Arsenault emphasized that the lithium-ion battery value chain varies depending on the recycling process design and the intermediate materials. These differences often represent supply chain re-entry challenges, which automakers can rectify by working with other supply chain entities to ensure the quality of intermediate materials meets the standards of re-entry. In closing, Arsenault remarked that emerging environmental, social, and corporate governance considerations in the EV industry have contributed to a global shift in the battery manufacturing and recycling processes.

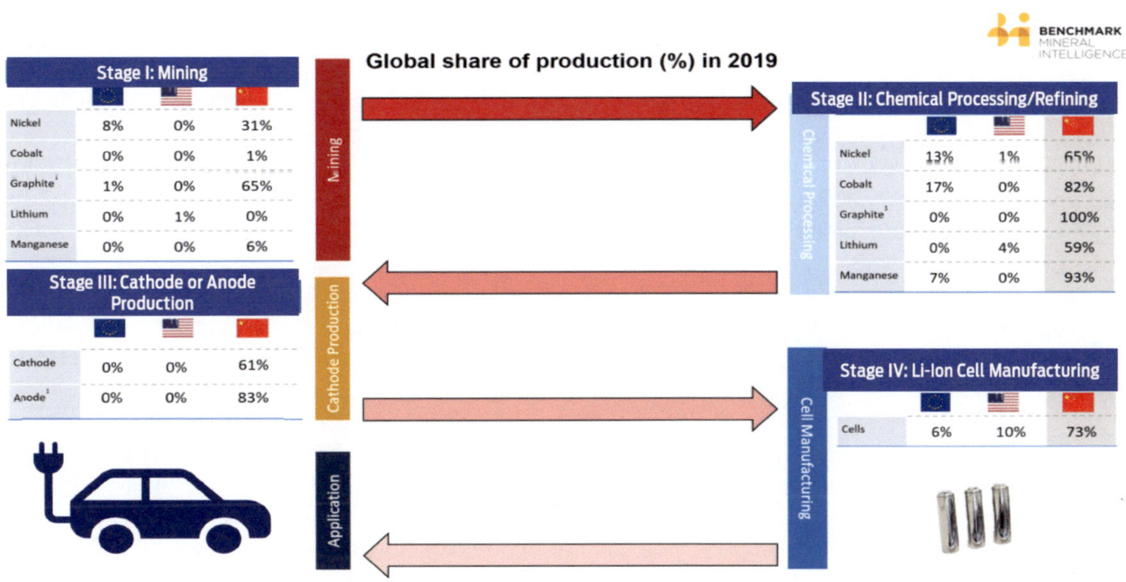

FIGURE 9 Flow chart showing each step of the lithium-based battery supply chain and the 2019 share of production for each stage in the European Union, the United States, and China. SOURCE: Renata Arsenault, Ford Motor Co., "Panel Discussion: Policy, Technical, and Market Options for Supply Chain and End-of-Life," presentation to the workshop, October 26, 2021.

Remarks by **John Graham**, professor of risk analysis at Indiana University, highlighted discrepancies between national government and local community interests related to acquiring raw materials for EV production. He illustrated this point using the example of lithium, first outlining why lithium mining and processing requirements will remain high through at least 2030: (1) EV battery chemistries will continue using significant amounts of lithium; (2) recycling lithium-ion batteries is currently not commercially viable; and (3) long vehicle lifetimes mean that there will be little lithium available for recycling and reuse. At the same time, lithium supply is not expanding rapidly enough to meet demand, in large part due to local opposition in areas with accessible lithium that results from concerns about stress on water systems, water quality, and impacts on other local businesses such as farms, ranches, and tourist operations. Such community opposition lengthens the permitting timeframe for lithium mining and processing. Graham further added that federal governments are not structured to address these issues; as an example, DOE is the lead agency to handle lithium supply, but permitting decisions reside with the U.S. Department of the Interior, the Bureau of Land Management, and the U.S. Fish and Wildlife Service. Consequently, lithium prices have soared to record 3-year highs in 2021, which, as Graham pointed out, could make meeting DOE's $60 per kilowatt-hour (kWh) goal difficult, but may increase the profitability of recycling. To conclude, Graham emphasized that similar price increases are seen with other materials for lithium-ion batteries and electric motors, which will make it more difficult for EVs to reach cost-parity with ICE vehicles and could slow consumer adoption.

CHALLENGES AND SOLUTIONS FOR THE ELECTRIC SYSTEM

To begin, **Chris Nelder**, creator and host of the Energy Transition Show podcast, discussed the complexities and challenges related to the grid side of EV integration. He indicated that the load requirement for EV charging can be very significant depending on the size and number of chargers. Direct current (DC) fast charging is a whole different game, Nelder added, and the industry is currently deploying DC chargers that exceed 150 kW. A charging station with six 150 kW DC chargers is equivalent to a load of a high-rise office building. This amount of load will have a significant impact on the grid and will require substantial investment in distribution system upgrades, Nelder explained. Utilities would require years of advanced planning to be able to accommodate the load increase associated with the EV transition. Additionally, Nelder noted that commercial fleet managers also need to prepare for the coming wave of vehicle electrification by equipping their yards and depots with charging stations. Nelder described two main techniques for effective EV load management: passive techniques like the design of utility tariffs for peak and off-peak charging and active techniques like using aggregators to control chargers and utilities controlling charges. On this note, Nelder hinted that the redesign of the grid is especially important to accommodate DC fast charging in areas with high loads. In addition, novel and innovative rate designs are critical to making the business model of DC fast charging more profitable. Permitting hurdles and poor guidance from utilities represent a major challenge and often cause the deployment of EV charging to be prohibitively expensive. Nelder concluded by remarking that deployment of EV charging is a complex problem that needs to be tackled at the legislative and utility levels.

A presentation by **Matthew Cloud**, lead EV engineer at National Grid, explored the grid impacts of EV fleets. In opening, Cloud stated that the push to accelerate vehicle electrification is a key strategy to decarbonize the U.S. economy. Due to commitments and mandates to decarbonize the transportation sector, fleet electrification is also expected to accelerate rapidly. Cloud noted that fleets are very suitable for electrification due to their high utilization rate and predictable travel distance. Moreover, electrified fleets can substantially improve air quality in disadvantaged communities. Citing a study conducted by National Grid,[9] Cloud explained that the full electrification of fleets in clustered locations could increase peak load on some circuits by more than 300 percent. The support for these electrical loads will require additional transmission and distribution solutions. In addition, seasonality can severely impact the peak load of the grid due to charging efficiency. Cloud suggested that minimum charging strategies that optimize charging utilization could significantly reduce grid impacts and the impact of fleet cluster electrification at the substation levels. Cloud closed by stating that long-term system

[9] National Grid and Hitachi ABB Power Grids, 2021, "The Road to Transportation Decarbonization: Understanding Grid Impacts of Electric Fleets," September, https://www.nationalgridus.com/media/pdfs/microsites/ev-fleet-program/understandinggridimpactsofelectricfleets.pdf.

planning and collaboration among stakeholders are critical to unlocking the potential for fleet electrification and reducing significant grid impacts.

Rohan Patel, senior global director of public policy and business development at Tesla, began by emphasizing the notion that EV charging can increase overall grid-system utilization and drive down rates through kWh billing. Rate design will be critical for the long-term success of utilities providing services for residential and non-residential EV charging users. Patel noted the importance of managing the non-coincident load demand for EVs to sustain the charging infrastructure by ensuring the EV load factor does not lead to higher energy costs. The soft costs associated with permitting and planning of EV charging stations are a major impediment, Patel observed. Volumetric time-of-use rates can be used to encourage good charging behavior, while gamifying and using simple tools can be an effective strategy to manage and encourage appropriate time-of-use practices. In closing, Patel stated that solving the current grid integration challenges associated with energy-producing assets, such as solar and wind, can provide an important roadmap for vehicle-grid integration (VGI).

Michael Kintner-Meyer, research engineer and systems analyst at Pacific Northwest National Laboratory (PNNL), framed his remarks by illustrating the complexity associated with EV grid integration from an electricity supply perspective. Using a back-of-envelope calculation, Kintner-Meyer showed that electric energy requirements to meet the current annual miles for light-, medium-, and heavy-duty vehicles are about 60 percent of the current electricity generation capacity in the United States. The power requirements for charging all on-road vehicles, assuming perfect load management, would be about 260 gigawatts of fully dispatchable generation. Kintner-Meyer noted that these are big numbers that will translate into huge grid investment needs. He then introduced and demonstrated the application of a distribution system analysis tool developed by PNNL with a case study in West Hollywood, California. This tool can help identify constrained components and location-specific bottlenecks in the grid while also supporting mitigation strategies to overcome bottlenecks. The PNNL team aims to further develop the tool and make it readily available to all utility and grid planners.

Jeremy Michalek, professor of engineering and public policy, professor of mechanical engineering, and director of the Vehicle Electrification Group at Carnegie Mellon University, opened by emphasizing that U.S. public charging infrastructure is lagging behind that of other major countries. He noted that limited household parking is a major impediment to BEV adoption in the United States. Without major infrastructure changes, the complete EV fleet transition is likely unrealistic. Additionally, high-speed public chargers will play a crucial role in the EV transition. However, even the highest speed charger may create some level of inconvenience (i.e., long queues) at service stations. Priorities to meet the public infrastructure needs include deploying high-speed charging along interstates and in neighborhoods where households depend on public chargers. Michalek then discussed EV charging from a life-cycle assessment (LCA) perspective. The literature on the impacts of EV charging can be confusing and can lead to cherry-picking data, Michalek observed. The confusion stems from the differences in fundamental questions and scopes of the research domain. Michalek explained that there are two main types of LCA questions: attributional and consequential. Attributional LCA research addresses the question of what emissions an EV is associated with or responsible for, while consequential LCA addresses the question of how emissions would change with greater EV adoption or EV policy. Michalek stated that consequential emissions are often fraught with high uncertainty due to potential changes in grid emissions over time, which is why some modelers prefer to use attributional methods and average emission factors. Depending on what question is being asked, it would be appropriate to either attempt to model and characterize the uncertainty rather than ignore the uncertainty, Michalek asserted. Furthermore, the relative consequential GHG benefits of EVs vary widely and depend on vehicle characteristics, regional grid mix, driving pattern, and climate. Michalek added that, generally speaking, vehicle-grid and utility-controlled charging can reduce costs of electricity generation, yet in certain regions, the consequential emission damage of coal plants can outweigh generation savings. In closing, Michalek remarked that coal retirement is central to making sure that vehicle electrification is in society's best interest.

Yamen Nanne, manager of distribution system development at the City of Los Angeles Department of Water and Power (LADWP), discussed the ongoing efforts of LADWP to prepare its electric grid for the onslaught of EV charging in the Los Angeles service territory. To begin, Nanne discussed the aggressive local and state-level policies and EV adoption targets driving transportation electrification in the city of Los Angeles. Representing about 10 percent of the state population, Los Angeles has set a goal of 750,000 light-duty EVs by 2030, which would require converting about 25 percent of the LDVs in the city. To accommodate that level of EV adoption, Nanne stated that the city would need about 120,000 charging stations in its service territory, including 3,000 fast chargers by 2030. Citing the LA100 Study conducted by LADWP and the National Renewable Energy Laboratory (NREL),[10] Nanne explained that electrifying about 80 percent of LDVs by 2050 would increase peak load by about 900-2700 megawatts (MW). He continued that LADWP currently offers EV commercial charging rate options to help encourage charging during off-peak and mid-peak hours. Nonetheless, with higher penetration of renewables and EVs, LADWP would have to redesign the time-of-use rates to align peak load EV charging with the peak supply of renewable electricity. Nanne noted that the LA100 identified Level 1 and 2 charging as significant contributors to the load in the year 2045, amounting to 2000-3000 MW (see Figure 10). Access to workplace charging infrastructure can help better align renewable generation with EV charging, Nanne concluded. In response to a question about alternative charging scenarios especially for dense urban areas, Nanne also described deploying curbside charging on light poles as well as sidewalk-placed chargers, which were some of the most utilized chargers in their service area.

Deepak Divan, professor and director of the Center for Distributed Energy at the Georgia Institute of Technology, elaborated on the challenges and opportunities of EV charging infrastructure. He noted that the transformation to a low-carbon economy cannot be achieved with 20th-century grid technologies alone. Divan believes that the poor efficiency of the ICE (using only 21 percent of the input energy), and the rapid emergence of lower-cost, higher-performance EV alternatives make the transition to a more efficient and carbon-neutral transportation sector unavoidable. A 60 percent year-over-year growth of EVs is starting to be observed, which implies a mid-level estimate of about 125 million EVs by 2040. Divan added that the rapid growth of transportation electrification will place enormous stress on the grid, drawing attention to the challenges of fast charging infrastructure in particular, such as cost-effective deployment, the possibility of poor utilization, challenges of grid access and peak stress on the grid, and viable business models. Divan then turned the discussion toward the utility level, where EV integration will pose some operational challenges; for example, changes resulting from EV integration can disrupt the business model of the utility, which often operates in a regulated and risk-averse environment. Divan underscored the fact

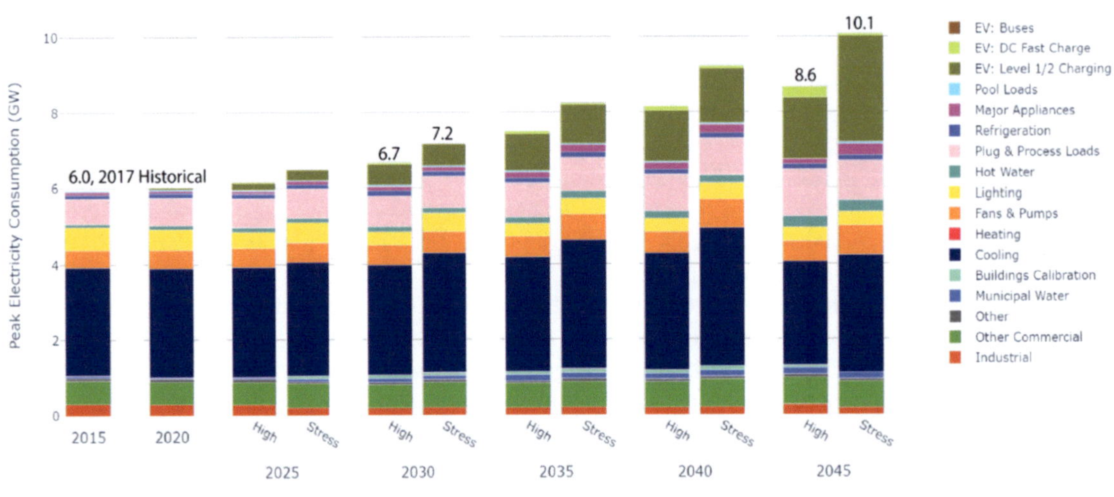

FIGURE 10 Peak electricity consumption by sector and scenario from 2015-2045. The "high" scenario indicates 60 percent access to residential charging and 50 percent access to workplace charging; the "stress" scenario indicates 90 percent access to residential charging and 15 percent access to workplace charging. SOURCE: Yamen Nanne, City of Los Angeles Department of Water and Power, "Electricity Generation and Load Profiles for EV Charging, and Related Impacts," presentation to the workshop, October 27, 2021.

[10] J. Cochran and P. Denholm, eds., 2021, *The Los Angeles 100% Renewable Energy Study*, NREL/TP-6A20-79444, Golden, CO: National Renewable Energy Laboratory, https://maps.nrel.gov/la100.

that tackling these grid integration challenges will require a holistic approach to address problems of sustainability, reliability, resiliency, affordability, and life-cycle cost. Decentralized generation and distributed energy resources for resiliency and grid services can offer a unique opportunity to serve the transportation needs. These new grid capabilities are currently being deployed in isolation and integrating these functionalities can be cost-effective. For instance, the use of grid-connected EVs for grid resiliency and grid support could pay for charging infrastructure. Divan concluded that the grid is on the cusp of fundamental transformation, and many of the elements are not under the control of the industry, which is why a holistic approach is needed.

Eric Cutter, senior director at E3, presented on grid distribution planning as a potential enabler or constraint to high vehicle electrification. He started by distinguishing between primary and secondary costs of grid distribution. Most studies on the cost and impact of vehicle electrification have been focused on the primary cost, Cutter explained. The primary cost is dominantly driven by the coincidence peak load at the sub-station level, while the secondary cost occurs at the final distributions line of the grid. The primary cost typically ranges between $20-170/kWh compared to the secondary cost of under $5/kWh. Referencing a national study on the potential distribution of high electrification loads, Cutter explained that the connected load for EVs can be significant due to the large magnitude of EV charging, potentially leading to significant investment costs. Currently, utilities are able to manage the coincident peak load to reduce the primary distribution cost, but they have limited capability in managing the secondary cost; this stems from the unwillingness of conservative engineers to adopt strategies and concerns about the disruptions high electrification can cause on the distribution system. Cutter also highlighted affordability as another concern, pointing to the aforementioned study which shows that adding all the EV load with an appropriate load management system can reduce the distribution rates by about 5 percent. The distribution system will be a key enabler or inhibitor for electrification.

Karen Glitman, senior director of distributed energy resources at the Center for Sustainable Energy, focused her presentation on the load profiles for EV charging and their related impacts on electricity generation. She explained that the LDV sector is the single largest contributor to GHG emissions in the United States, and load management of EVs will play a key role in the efficient adoption of EVs. Glitman added that the flexible load cycle of LDVs presents an interesting opportunity to consider them as mobile appliances or batteries on wheels. The levers for controlling EV loads, she continued, can be achieved through direct or indirect means like time-of-use rates. Additionally, EVs have the built-in capability to respond to utility signals that control and shift utilization and the charging rate. Glitman believes these capabilities can serve as a useful tool and asset for grid managers. In terms of market opportunities, Glitman explained that EV charging can serve as a revenue stream for property owners as deployment and charge requirements reach substantial levels. Homeowners associations, fleets, airports, and car rentals can all tap into this market opportunity. Glitman closed by emphasizing time-of-use, location, and load size as important considerations to ensure an efficient, expeditious, and equitable transition to a clean transportation future.

Jason Hills, manager of distribution and electrification at LADWP, started his presentation by recognizing LADWP's successful installation of more than 10,000 commercial charging stations in the city of Los Angeles—ahead of the city's 2022 target. With Governor Newsom's new EO banning the sale of gas power vehicles by 2035,[11] the city of Los Angeles needs to reach more aggressive milestones. The city will need about 120,000 charging stations by 2030, Hills added. Citing the LA100 study conducted by LADWP and the NREL,[12] Hills then explained that an 80 percent adoption of light-duty EVs by 2045 will significantly impact the electric system load profile. The LADWP distribution planning and development team are taking proactive measures to accommodate the load growth from EV charging in the near term. Based on the most recent adoption trends and forecasts, LADWP anticipates a fivefold increase in peak load over the next 10 years. Most of this growth will occur on 34.5kV systems due to this system's ability to handle larger and higher power capacities than one can expect from medium- and heavy-duty vehicles. To handle the anticipated load from EVs, LADWP is also upgrading its

[11] Newsom, 2020, "Executive Order N-79-20," Executive Department, State of California, September 23, https://www.gov.ca.gov/wp-content/uploads/2020/09/9.23.20-EO-N-79-20-Climate.pdf.

[12] J. Cochran and P. Denholm, eds., 2021, *The Los Angeles 100% Renewable Energy Study*, NREL/TP-6A20-79444, Golden, CO: National Renewable Energy Laboratory, https://maps.nrel.gov/la100.

distribution system with new stations, voltage conversion, and automation. At the substation level, the distributional planning team anticipates major upgrades to the receiving and distribution stations by 2035. As part of its grid modernization efforts, LADWP is looking to convert portions of the 4.8kV distribution to a higher voltage, a process that will help reduce losses, increase capacity for loads, and enable the hosting of higher distributed energy resources. In addition to upgrading the core grid infrastructure, Hills said, LADWP is also exploring cost-effective load management programs to accommodate the vehicle to grid transition.

Sara Baldwin, director of electrification policy at Energy Innovation, summarized findings from a report[13] focusing on the impacts of transportation electrification on consumers, emissions, public health, employment, and the electric grid. The study explored two comparative scenarios: a drive clean scenario and a no new policy scenario (see Figure 11). Baldwin explained that the drive clean scenario assumes EVs as 100 percent of new LDV sales by 2030, EVs as 100 percent of new medium- and heavy-duty vehicle sales by 2035, and a 90 percent clean electricity grid by 2035. In the "Drive Clean" scenario, there would be a substantial ramp-up of EV adoption compared to the business-as-usual, "No New Policy" case. Key findings from the report show that accelerating EVs and a clean grid can lead to significant benefits for the economy, the average consumer, and the environment. These benefits include consumer savings of about $2.7 trillion between 2020 and 2030, the creation of some 2 million jobs by 2035, and a 96 percent reduction in pollution-related premature deaths by 2050. Referring again to the study, Baldwin underscored the finding that a 90 percent clean grid can indeed handle the additional demand from the drive clean scenario. Consistently, the results also indicated that the grid can support a 90 percent clean energy portfolio without adverse impacts on overall reliability. As the costs of renewables and storage continue to decline, the drive clean scenario results suggest that the average cost may decline for the average ratepayer. Highlighting relevant insights from a companion policy report, Baldwin explained that many of the policy and regulatory pathways considered in the report are either under way or under discussion in some form across states.

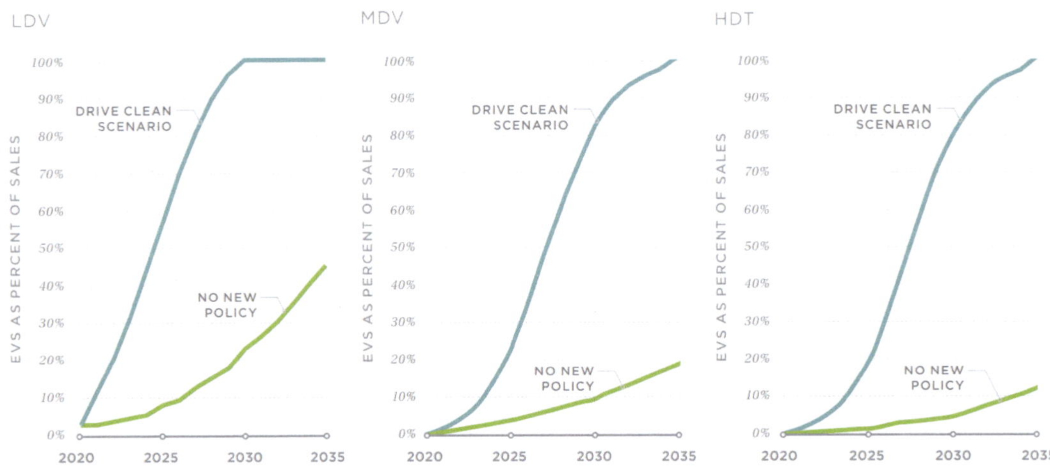

FIGURE 11 Electric vehicles (EVs) as a percentage of sales, comparing the drive clean scenario in blue with the no new policy scenario in green. NOTE: HDT, heavy-duty truck; LDV, light-duty vehicle; MDV, medium-duty vehicle. SOURCES: Sara Baldwin, Energy Innovation, "Accelerating EVs and a Clean Grid: Impacts and Policy Considerations," presentation to the workshop, October 27, 2021; from A. Phadke, N. Abhyankar, J. Kersey, T. McNair, U. Paliwal, D. Wooley, O. Ashmoore, et al., 2021, *2035 Report: Plummeting Costs and Dramatic Improvements in Batteries Can Accelerate Our Clean Transportation Future*, April, https://www.2035report.com/transportation.

[13] Energy Innovation, Goldman School of Public Policy, and GridLab, 2021, 2035 Report 2.0: *Plummeting Costs and Dramatic Improvements in Batteries Can Accelerate Our Clean Transportation Future*, April, http://www.2035report.com/transportation/wp-content/uploads/2020/05/2035Report2.0-1.pdf?hsCtaTracking=544e8e73-752a-40ee-b3a5-90e28d5f2e18%7C81c0077a-d01d-45b9-a338-fcaef78a20e7.

Maria Bocanegra, commissioner with the Illinois Commerce Commission, discussed the regulatory opportunities and challenges for massive EV adoption and grid impacts from a policy perspective. She emphasized that incentives represent an opportunity for regulators to review and manage grid investments for vehicle integration. As an example, tailored qualifying infrastructure or distribution improvement charges can be used to incentivize utilities by providing a guaranteed return on investments through surcharge rates. Bocanegra cautioned that these investments ought to be prudent, reasonable, and at the least cost. Another policy opportunity lies in the potential for data generation and access that comes with the ability to communicate and interact with grid operators and EV consumers. Bocanegra added that telemetric technologies that capture data also present an opportunity for regulators in determining whether utilities are managing their grid as effectively and efficiently as possible. Moreover, data access allows utilities to address system operational challenges and for strategic investment planning. Bocanegra highlighted several policy opportunities, including interagency coordination and addressing ownership of and access to behind-the-meter assets, especially when pairing charging with renewable sources. She noted that memoranda of understanding are beginning to emerge among utilities around electrification to establish national charging station corridors. Bocanegra then drew attention to some of the major policy impediments, asserting for example that the balancing act between affordable rates and ensuring that utility investments result in reliable services is an important challenge for regulators. Additionally, regulations that lag behind expected policy formulation and adoption can pose key policy challenges as well, Bocanegra concluded.

Daniel Bowermaster, senior program manager for Electric Transportation at the Electric Power Research Institute (EPRI), opened by highlighting that in the United States, the EV sales increase for 2021 exceeded that of the overall automotive market. The 3.2 percent market share of EVs in the United States is impressive, especially considering the 16-17-year automotive fleet turnover rate. Local and national policies, coupled with the automotive industry's strong reactions to the policies, underpin this major upward trend of EV sales, Bowermaster explained. Strong environmental and economic factors drive the sales of EVs in China and the European Union; in the United States, Bowermaster suggested that a clearer picture of EV sales may emerge in the next few years as the EV industry enters the market segments for trucks and sport utility vehicles (SUVs). Bowermaster then referenced an EPRI chart showing some of the notable EV penetration trends in various U.S. counties to illustrate the importance of ZEV regulation in ramping up EV market share (see Figure 12). Furthermore, mass adoption

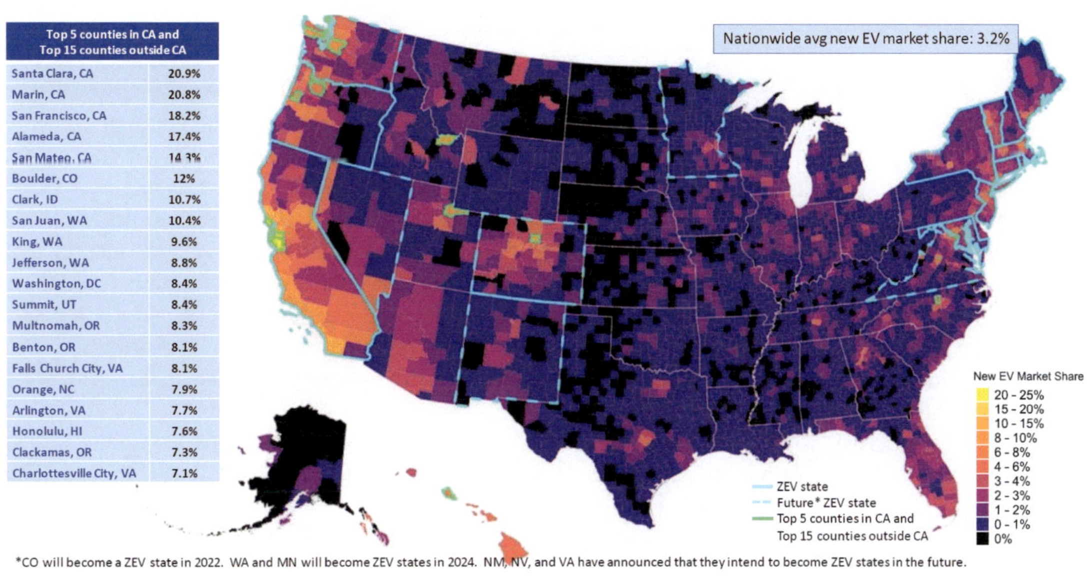

FIGURE 12 New EV market share by U.S. county from September 2020-August 2021, where a lighter color indicates a higher share. States with a ZEV program are outlined in solid blue, future ZEV states in dotted blue, and the top five counties in California and top 15 outside California in solid green. SOURCE: Daniel Bowermaster, Electric Power Research Institute (EPRI), "Electric Transportation: State of the Market and Preparing for the Future," presentation to the workshop, October 27, 2021.

of EVs is also occurring in states that do not necessarily have the regulatory framework in place despite perhaps having a highly competitive market, Bowermaster added. While utilities are taking joint actions to prepare for EVs, Bowermaster believes there is still a long way to go and many lessons to learn about deploying easily accessible and reliable public charging infrastructure.

Alan Jenn, assistant professional researcher at the Plug-In Hybrid and Electric Vehicle Group within the Institute of Transportation Studies at UC Davis, discussed the electricity system needs for EV integration from a policy, technical, and market option perspective. Jenn began his presentation by noting that the focus of most practitioners is to get EVs on the road and secure adequate EV infrastructure. Therefore, the development of policy in this area will be crucial to mitigate the potential challenge of VGI. One of the major challenges, Jenn observed, is the whole host of different stakeholders in the landscape of grid integration. He then acknowledged that there is a large body of policies paving the way for VGI in certain parts of the country. For example, several regulatory agencies in California have come together to develop a framework for key VGI policy avenues, including the technology infrastructure, market mechanism, and adoption. Jenn added that the many potential programs and initiatives for each policy avenue are necessary to developing a robust VGI ecosystem. Nevertheless, a key question that needs to be addressed is whether the load and load management for EV charging will be accommodated on the distribution or wholesale electricity system. Additionally, evaluating the value of EV charging across both systems and how to design feedback mechanisms for users to reap the benefits will play a key role moving forward. In his presentation and in response to a question about vehicle to grid applications, Jenn explained that much of the current focus has been on the technological implementation and feasibility of VGI; however, more work is needed to understand the intersection of these technological advances with the way users interact with VGI mechanisms and schemes. In his closing remarks, Jenn stressed that better coordination is needed to facilitate lessons learned between pilot programs.

MEETING CONSUMER NEEDS

Tyson Jominy, vice president of data and analytics at J.D. Power, discussed the market dynamics of EV sales in the United States. He explained that the U.S. EV industry can be described in three distinct evolutions: EV 1.0, EV 2.0, and EV 3.0. The first evolution of EVs was characterized by the adaptation of existing no-frill hatchbacks aimed to target the lower end of the consumer market. The low-price target came with compromises on battery technology and vehicle range. In fact, most of the models in the EV 1.0 class have been discontinued by the manufacturer due to their abysmal market performance, Jominy said. He described the current EV market trend (i.e., EV 2.0) as a phase marked with premium brand electric powertrains with a higher technology performance and range. The higher performances of EV 2.0 vehicles come at a higher cost to the consumer. Automakers in the EV 2.0 market have created dedicated platforms for EV and unique features for flagship brands, Jominy added. For EV to become mainstream for the average consumer, the industry has to evolve to the EV 3.0 phase. Currently, EVs are almost absent in the seven largest automobile segments. Targeting these large market segments is critical for EVs to gain traction; for instance, electric SUV models are currently transforming the SUV market by outselling other non-electric SUVs. Targeting the mainstream markets such as the SUV and truck market can help increase the automaker's profit margins and reduce the need for extreme incentives. In closing, Jominy noted that once EVs achieve full acceptance in the mainstream markets, the industry can then focus on making EVs accessible to less affluent consumers.

Ahmed Abdulla, assistant professor in the Department of Mechanical and Aerospace Engineering at Carleton University, provided insight into the underlying drivers and behavioral trends of EV adoption and charging in southern California. Most of these insights were based on a survey of 643 EV owners recruited from UC San Diego, UC Irvine, and the EV Club of San Diego, a sample population that, Abdulla noted, skewed more male, more liberal, wealthier, and more educated than the average person living in the United States. He cautioned that due to limitations of the sample, any comments about the adoption and charging trends should be couched in these caveats. The study aimed to address two issues: (1) changes in the adoption profile and attitudes toward EV incentives and (2) charging behavior and how users interact with workplace charging. The survey was novel because actual EV owners could provide unique insights into the underlying roots of their decisions, and because eliciting the preferences of owners across several years could reveal evolutions in behavior or in attitudes toward EV incentives and workplace charging. The study revealed two major groups of EV users based on how satisfied they are with the incentive they received. The first group—EV owners that were more satisfied with incentives—were generally wealthier and had shorter commutes. On the other hand, people who disliked their incentives had longer commutes, a lower preference for environmental protection, and lower levels of income and education. This insight raises important questions about whether to maintain incentives or modify them as the adopter profile changes. To put it bluntly, Abdulla added, incentives should probably be maintained or increased instead of reduced, since the people who were unsatisfied with incentives skewed more toward mainstream America. Survey results on charging behavior revealed that most university campus users plug into the chargers in the morning as soon as they get to campus. Notably, the mode of campus chargers occurs between 7 and 8 am, which does not align with the California peak for solar power production. The survey found that commuters who more recently adopted EVs were more often the ones charging earlier than the maximal peak for solar power production. Abdulla added that these charging challenges might change as infrastructure is deployed, but they are not going away in the short term and need to be managed through demand- and supply-side interventions. The results also revealed some emerging opposition in terms of risk that can stymie adoption or cause a plateau. For example, cybersecurity and data privacy concerns were voiced by some later EV adopters. The EV industry must anticipate and address these risks. As the EV revolution unfolds, identifying emerging challenges and addressing them preemptively will be critical. Moreover, the more important policy implication for infrastructure developers is that relying on operating experience, even from the recent past, may not be the soundest course of action for developing EV infrastructure.

Shelley Francis, co-founder and principal at EVNoire, discussed the importance of centering equity in the e-mobility policy discourse. She began by asserting that mobility is about more than just getting from point A to point B. Rather, mobility can be described as providing access to health centers, education, employment, economic development, and environmental quality. Francis explained that over the past hundred years, car ownership has been one of the most powerful economic drivers and wealth builders for U.S. families. She then illustrated the historical connection between segregation and unequal mobility for Black Americans, explaining that destructive transportation policies and practices tend to skew toward Black American and Latin Heritage communities. These policies and practices have disproportionately exposed minority communities to detrimental public health impacts. The transition to e-mobility presents a unique opportunity to prevent past transportation inequities, Francis stressed. A human-centric approach of engaging and partnering with communities can be an effective tool to avoid the foreseeable inequities of e-mobility. Engaging diverse communities involves listening to better understand the solutions and barriers and also empowering communities to become more involved. Francis also emphasized, "When you engage with communities, one of the critical things to understand is that things will move at the speed of trust." She concluded by remarking that effective engagement is a continual and long-term process of nurturing relationships. Responding to an audience question about improving access to EVs in disadvantaged communities, Francis emphasized the needs to (1) allocate funds for research on engaging and educating diverse communities about EV technologies; (2) make rebates and incentives accessible and available to all consumers; and (3) highlight all potential benefits of e-mobility, including workforce and economic opportunities.

Gil Tal, director of the Plug-In Hybrid and Electric Vehicle Group within the Institute of Transportation Studies at UC Davis, opened by describing the three equity goals of vehicle electrification: make EVs available to all buyers, share public funding, and secure environmental benefits. Tal then discussed the 2016-2019 trend of new LDV sales in California, pointing to the fact that most residents of the state are not in the new car market. The trend in California also indicates low-income households will start adopting used EVs in less than 10 years. Tal added that the market segment of low-income consumers would contribute substantially to California's 100 percent EV target. Currently, the adoption of both new and used PEVs is skewed toward non-disadvantaged communities in California. Notable trends in California indicate that there is limited access to public charging for multi-unit dwellings in minority and low-income communities. In comparing the advantages and disadvantages of used PEVs, Tal explained that for low-income buyers, used PEVs can present substantial economic benefits in operation and maintenance. Nevertheless, the drawbacks associated with EVs' initial cost, reliability, and access to charging can present a risk to low-income buyers. In closing, Tal summarized several relevant policy discussions and initiatives that can help bridge the adoption gap for low-income consumers, including subsidizing and installing chargers, controlling charging cost, encouraging high vehicle turnover, requiring and subsidizing battery warranties, subsidizing PEVs for low-income users, and incentives aimed at closing total cost of ownership gaps.

Zhenhong Lin, senior research and development (R&D) staff member at the National Transportation Research Center of Oak Ridge National Laboratory, explored the implications of free charging for low-income communities. He began with a general overview of the global EV trend, noting that Europe and China have surpassed the United States to become the market leaders in EV adoption. While the penetration of EVs is not sufficient to meet climate goals, the current rate of EV adoption is remarkable compared to other alternative fuel vehicles. Lin suggested that the adoption of EVs is not equitable and varies across gender, income, education, and other demographic factors. He added that the inequities of EV adoption extend beyond demography and include access to incentives and charging infrastructure. Due to these inequities, low-income consumers are missing out on the benefits of vehicle electrification. Lin then explained the notion of free charging coupled with affordable used PEVs for low-income communities. He argued that free charging, regardless of the funding mechanism, presents a net positive value to the society, similar to other public systems such as the public road system. He argued that affordable used PEVs, if adopted by low-income communities due to the stimulus of free charging, can reinforce the new PEV demand and therefore improve both effectiveness and equity of vehicle electrification. Lin closed his presentation with a discussion on the broad implications and open research questions that need to be addressed in the context of freely charged used PEVs. One example, highlighted in the question and answer session, is the overall environmental impact of free charging on emissions and local air pollution. On the one hand, the replacement of ICEVs with EVs, coupled with the increasing share of clean electricity on the grid, could yield positive air quality benefits. However, free charging may incentivize driving, thereby increasing energy use associated with EVs and shifting emissions upstream, which could exacerbate local air pollution near electricity generation facilities. Lin emphasized that more quantitative and location-specific analysis would be required to make any conclusions about these potential impacts.

Matthew Metz, founder and co-executive director of Coltura, explained that targeting EV incentives and policies on the biggest users of gasoline can serve as a strategy to reduce U.S. personal transportation emissions. "Gasoline superusers" are defined as the top 10 percent of drivers in terms of gasoline consumption. They account for one-third of all gasoline burned. On average, superusers drive about 30,000 miles per year, which is about three times more than the average U.S. driver. Additionally, superusers tend to drive bigger vehicles like pickups and SUVs. With regard to household income, superusers are similar to the general population of drivers—clustered mostly in the lower- and middle-income levels (unlike EV drivers, who skew toward the highest income levels). Superusers tend to spend more of their total annual income on gasoline than the average American driver; the gasoline cost burden is even steeper among lower-income superusers. Metz suggested that if low-income superusers make the switch to EVs, they would save hundreds of dollars per month that would otherwise be spent on gasoline. As illustrated in Figure 13, designing incentives based on historical gasoline consumption, such that the biggest gasoline users receive the biggest incentives, and focusing EV education efforts in geographic locations with high concentrations of superusers are two promising strategies that could help increase EV adoption among superusers.

A presentation by **Nicholas Millar**, product manager at EV Connect, described how charging networks can help meet the needs of EV drivers. Due to the ubiquity of electric outlets, EV charging represents a more convenient fueling experience relative to ICE vehicle fueling, Millar opened. However, non-standardized connectors, lack of universal access to overnight charging, and closed networks are major barriers to EV charging access. In light of these challenges, EV providers have recently made efforts to allow roaming across networks that can simplify EV charging and overcome the closed network barrier. Another benefit, Millar added, is the potential for lower fueling costs for EV charging on these networks. It is worth noting that these cost savings might not be realized everywhere in the United States due to the high infrastructure upgrade cost and the lack of incentives to embrace the flexibility of EV charging. These can be overcome with smart design and smarter charging software. He noted that light-, medium-, and heavy-duty fleets can benefit the most from lower cost-per-mile. Furthermore, a more holistic charging solution for commercial fleets can have a spillover effect in speeding the charger coverage for non-fleet users.

Kelly Yearick, senior program manager at Forth, discussed the opportunities and challenges of transportation electrification in rural communities. She explained that transportation electrification opportunities in rural America

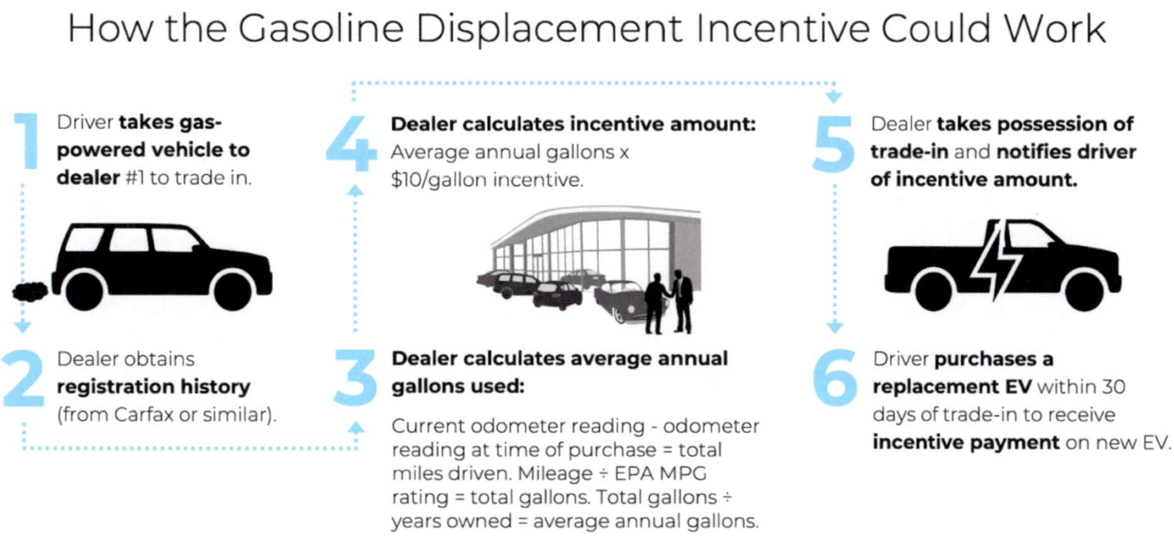

FIGURE 13 Illustration of a potential gasoline displacement incentive, in which a driver's annual mileage is correlated to the incentive payment received for an electric vehicle (EV) purchase, with higher mileage receiving a greater incentive. Such incentives could encourage the biggest users of gasoline to make the switch to EVs while saving money. SOURCE: Matthew Metz, Coltura, "Gasoline Superusers," presentation to the workshop, October 28, 2021.

are substantial. The opportunities lie in the fact most rural dwellers tend to drive long distances and own older, larger, and less efficient vehicles. Yearick added that rural dwellers are also more likely to live in single-family homes lending to more improved access to charging. However, transportation electrification in rural regions is not without challenges, which include less awareness of EVs and limited access to public transportation and mobility services. Yearick stressed that substantial incentives and outreach will be required to overcome the entry barrier for EV adoption in rural American given the lower average income and aging demography. She concluded that the lack of EV models suitable for rural consumers is a major barrier that needs to be addressed to increase rural transportation electrification. Asked about effective approaches to increase EV adoption in rural communities, Yearick noted that adoption of EVs by local governments and well-respected local organizations could both increase awareness of these vehicles in the community and incentivize charging infrastructure development in the area.

In her opening remarks, **Britta Gross**, managing director of the Carbon Free Mobility Global Program at RMI, expressed excitement about the upward trend in EV sales across states and the country as a whole. However, looking at the current EV sale trend from a top-down perspective indicates that the current pace of EV scaling is insufficient to meet the goal of climate mitigation, Gross explained. About one in four cars and one in seven trucks on the road must be electric to meet a 45 percent carbon emissions reduction in the transportation sector by 2030. Gross noted that the challenge to achieve this goal is staggering, and she believes the time to act is now since there is considerable momentum and pressure on the part of states, investors, automakers, and the government to increase the market share of EVs. The infrastructure needs are substantial, Gross added. For example, even in the optimistic and aggressive home charging scenario of 78 percent, the United States will still need about 375,000 DC fast chargers in 2030. To reach this goal, the pace of DC fast charging installation must be increased from the current 600 charging ports per quarter to 10,000. Moreover, the unprofitable economics of operating charging stations and deploying EV fleets present additional challenges that need to be resolved. Gross cited a case study of Los Angeles, explaining that ride-hailing services can be used to bridge the inequities of e-mobility and DC fast charging access. In her closing remarks, Gross highlighted the need to tackle systemic issues related to permitting, interconnection, the economics of charging, and equity for large-scale electrification.

Carla Bailo, president and chief executive officer (CEO) of the Center for Automotive Research, provided a summary of EV sales and demographics and offered a high-level perspective on the policies being negotiated.

She started her presentation by noting the substantial increase in the EV share of the LDV market, from 3.8 percent in 2020 to 8.9 percent in 2021. The growing acceptance of EVs is due to the reduction in driving need and realization of the importance of clean air during the COVID-19 pandemic. On a segment-by-segment basis, the increase in EV sales is staggering. Bailo added that product functionality and the availability of the different EV products will be critical for wider acceptance. Referring to EV sales data from California, Bailo then pointed to the fact that a sizeable share of new EV buyers has a household income of more than $100,000, which represents a significant gap in access to EVs by the middle class considering the average American income is about $65,000. To this point, Bailo noted that people are keeping their cars longer and longer vehicle turnover rates are another lever that must be considered when thinking about the future of electrification. She then drew attention to the factors causing the shift to electrification, many of which have been covered by previous speakers. Notably, for the first time, U.S. EV light vehicle sales are beginning to move independently of the real price of gasoline (see Figure 14). Additionally, production of EVs is expected to significantly increase by 2025 in line with announced plans being put in place by automakers as well as President Biden's EO on Strengthening American Leadership in Clean Cars and Trucks, which sets a goal of 50 percent sales of emission-free vehicles by 2030. Biden's EO also includes investment in charging infrastructure, domestic manufacturing, and consumer incentives aimed at the middle class. Even so, Bailo noted the provisions for strengthening U.S. R&D are very limited compared to other countries like China, and it is therefore imperative to develop technologies domestically by funding students and universities to conduct research. Moving forward, another important consideration is building resiliency into our supply chain of raw materials supply and critical technologies.

Yan (Joann) Zhou, principal analyst and group leader of the Mobility and Deployment Group at ANL, began by explaining that households in certain low-income census tracts have a transportation energy burden, spending more than 20 percent of their income on transportation. Zhou noted that an ANL analysis found fuel efficiency to be a primary contributor to transportation energy burden. For example, from 2016-2018, a 3 percent improvement in fuel efficiency saved American households more than $1 billion. Zhou added that increasing the adoption of high-efficiency vehicles such as EVs can improve energy equity, especially across low-income households. In collaboration with other national labs, Zhou and her colleagues at ANL recently completed a study on the com-

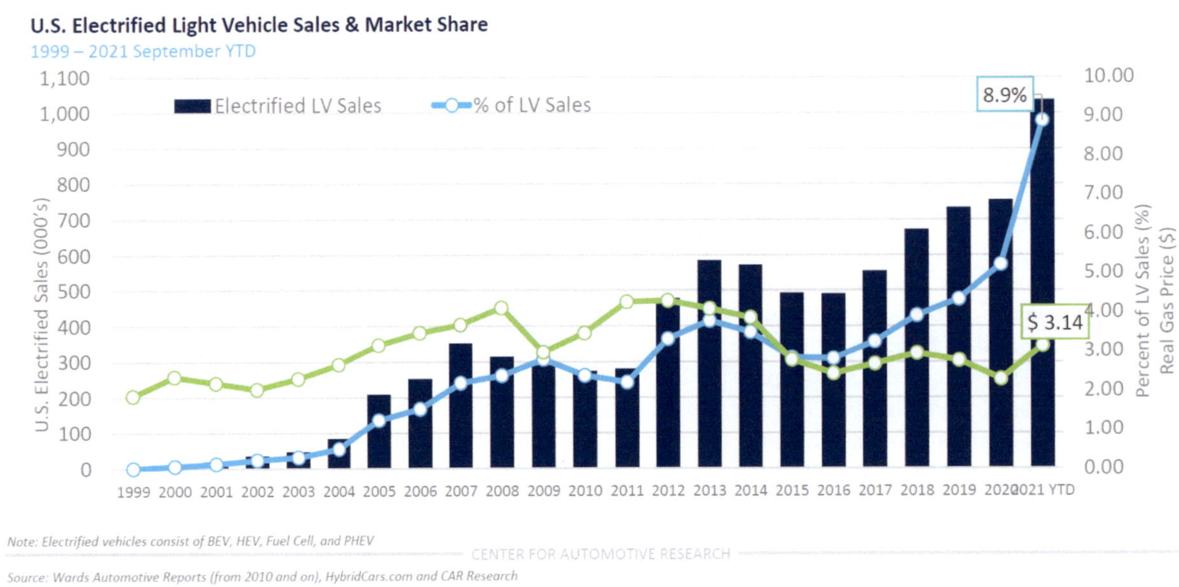

FIGURE 14 Plot of the number of electrified vehicle sales (navy blue bars) and electrified vehicle sales as a percent of all light-duty vehicle sales (blue line) in the United States from 1999-2021, compared to gasoline prices (green line) during the same time period. NOTE: LV, light vehicle; YTD, year to date. SOURCE: Carlo Bailo, Center for Automotive Research, "Electric Vehicles Sales and Policy Review," presentation to the workshop, October 28, 2021.

prehensive total cost of vehicle ownership, which identifies eight important cost components: vehicle, financing, fuel, insurance, maintenance and repair, tax and fees, payload changes, and labor. Looking at the model year 2025, the study results indicate that PHEVs will be comparable with conventional vehicles. In addition, the cost of BEVs with 300 mile range will be slightly higher than conventional vehicles, but still comparable due to lower maintenance and repair costs. Zhou stated that major vehicle cost components include purchase and depreciation; early EV models face higher depreciation costs and some early models are worth only $6,000 after 6 years of regular use. Additionally, while the retention rate and battery degradation are also considered major cost barriers for EV consumers, ANL analysis shows that the retention rate for newer models has greatly improved over the last few years. Moreover, mature powertrains in particular have improved consistently compared to their conventional competitors. Zhou added that with these improvements, one recognizes that early market models have created a low-cost EV market for low-income communities. In the future, deploying chargers more widely and strategically can capture about 90 percent of charging demand in the public domain.

Sandra Wappelhorst, senior researcher on the Europe team at the International Council on Clean Transportation, introduced her presentation by discussing the countries and subnational governments that have made sales commitments for new electric LDVs. Auto manufacturers are reacting to these policies by announcing their commitments to full EV production in the medium term. Referring to a map of countries and U.S. states that have made official commitments toward 100 percent new EV sales, Wappelhorst noted that these countries accounted for about 14 percent of total new global passenger car sales in 2020. For larger markets like the United States and Europe, commitments to EV targets will send an important market signal. In Europe, tighter CO_2 emissions standards from 2020 model year cars have generally been a key driver for spurring the electric passenger car market in European countries in 2020 and 2021 (see Figure 15). The development in European countries exemplifies the importance of regulations, Wappelhorst remarked. Furthermore, some leading car markets like Germany, France,

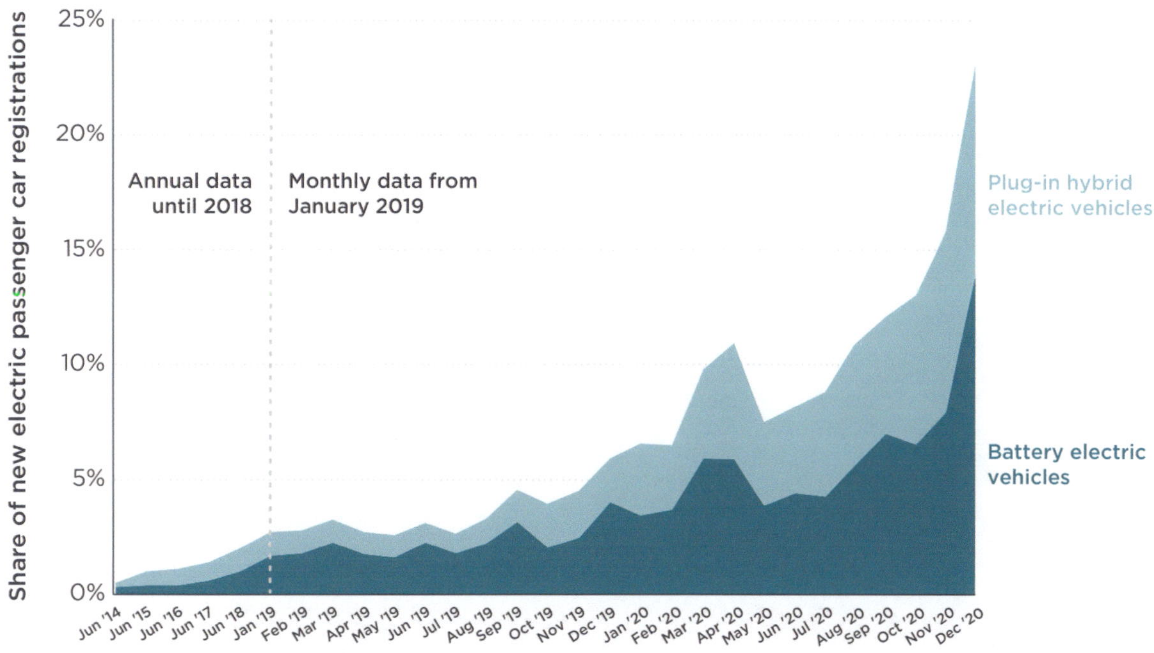

FIGURE 15 Share of new electric passenger car registrations in European Union member states and the United Kingdom from 2014 to 2020, showing significant increases upon the introduction of tighter CO_2 emissions standards in model year 2020. SOURCES: Sandra Wappelhorst, International Council on Clean Transportation, "Driving the Electric Vehicle Market: Lessons from Europe," presentation to the workshop, October 28, 2021; from S. Wappelhorst, 2021, "Europe's CO_2 Emission Performance Standards for New Passenger Cars: Lessons from 2020 and Future Prospects," International Council on Clean Transportation (ICCT), licensed under CC BY-SA 4.0, https://creativecommons.org/licenses/by-sa/4.0.

Spain, and Italy have introduced COVID-19 stimulus packages that include increased incentives for consumers to buy new EVs. It is worth noting, Wappelhorst stated, that the high registration share in Europe is also driven by the registration of company cars. For instance, in the Netherlands and Germany, more than 70 percent and almost 50 percent, respectively, of new electric cars in 2020 were registered by companies. Thus, company cars and the private use of company cars can spur the EV market in addition to the private car market. Wappelhorst then described the notable EV adoption trend in rural regions of Europe for the year 2019, particularly in Scandinavia, Portugal, Germany, Austria, and Switzerland, where in some cases EV uptake equaled or exceeded the 3.6 percent European average. Higher public charging density than the national average is one key factor driving the uptake of EVs in some rural regions of Europe. Wappelhorst offered the example of a rural region in Germany, which has shown success in increasing the awareness and adoption of EVs by establishing education facilities, companies that produced components for EVs, and car dealerships to handle the sale of EVs. To conclude, Wappelhorst emphasized that increasing the EV market share will require a complex system approach of regulation, incentives, charging infrastructure, and local action.

Following their presentations, the panelists in the final session—Gross, Bailo, Zhou, and Wappelhorst—discussed potential barriers to EV adoption that will remain even once these vehicles reach cost parity with ICE vehicles. Gross highlighted the numerous infrastructure challenges that need to be addressed, including location and number of charging stations, permitting and grid interconnect requirements, industry standards for chargers, and building codes for charging installations. Bailo and Zhou added the need for better consumer education on EV range and total cost of ownership, with Bailo noting that automakers have begun working with dealer networks to train salespeople to better communicate about EVs to potential buyers. Additional barriers mentioned were the impacts of weather conditions on battery life, model availability to meet all consumer needs, and material shortages and supply chain challenges that may arise with widespread EV adoption. With the caveat that vehicle markets vary widely across countries, Wappelhorst shared several lessons learned from EV adoption in Europe, emphasizing in particular the importance of model availability and charging infrastructure development. Increasing model availability allows EVs to reach a wider consumer base with different driving needs. Making charging infrastructure more convenient to the consumer is also critical for EV adoption, and can include deploying faster charging options, streamlining payment systems, and standardizing charging rates.

Appendixes

A

Workshop Agenda

DAY 1—OCTOBER 25, 2021
ELECTRIC VEHICLES 101 AND TRANSPORTATION DECARBONIZATION

1:00 PM ET **Introductory Remarks**
Steven Cliff, Acting Administrator, National Highway Traffic Safety Administration

1:10 PM ET **Welcome and Overview of the Electric Vehicle Landscape**
Gary Marchant, Arizona State University, moderator

1:30 PM ET **Technical Talks and Q&A:** Introduction to BEV Technologies
Nady Boules, NB Motors, LLC
Deidre Strand, Wildcat Discovery Technologies
Denise Gray, LG Energy Solution Michigan Inc. Tech Center
AK Srouji, Romeo Power

2:40 PM ET **Technical Talks and Q&A:** Introduction to BEV Consumer and Regulatory Considerations
Dave Cooke, Union of Concerned Scientists
Alexander Edwards, Strategic Vision
Kate Whitefoot, Carnegie Mellon University
Joshua Linn, University of Maryland

3:50 PM ET **Break**

4:00 PM ET **Technical Talks and Q&A:** Transportation Decarbonization
Rachael Nealer, White House Council on Environmental Quality
Patty Monahan, California Energy Commission
Daniel Sperling, University of California, Davis

5:00 PM ET Adjourn for the Day

DAY 2—OCTOBER 26, 2021
VEHICLE PRODUCTION AND LIFE CYCLE

1:00 PM ET **Welcome**
Anna Stefanopoulou, University of Michigan, moderator

1:05 PM ET **Technical Talks and Q&A:** Impacts to Existing Supply Chains
Kristin Dziczek, Center for Automotive Research
Brad Markell, AFL-CIO
Zoe Lipman, BlueGreen Alliance

2:00 PM ET **Technical Talks and Q&A:** New Supply Chain Needs Specific to BEVs
Warren Day, U.S. Geological Survey
David Klanecky, Piedmont Lithium
Celina Mikolajczak, QuantumScape

3:00 PM ET **Break**

3:15 PM ET **Technical Talks and Q&A:** Vehicle End-of-Life Options
Linda Gaines, Argonne National Laboratory
Fan Dai, University of California, Berkeley
Anna Stefanopoulou, University of Michigan

4:15 PM ET **Panel Discussion:** Policy, Technical, and Market Options for Supply Chain and End of Life
David Howell, DOE-Vehicle Technology Office
Renata Arsenault, Ford Motor Company
John Graham, Indiana University

5:15 PM ET Adjourn for the Day

DAY 3—OCTOBER 27, 2021
CHALLENGES AND SOLUTIONS FOR THE ELECTRIC SYSTEM

1:00 PM ET **Welcome**
Zhenhong Lin, Oak Ridge National Laboratory, moderator

1:05 PM ET **Technical Talks and Q&A:** Electricity Demand for Charging at Mass Market EV Deployment
Chris Nelder, The Energy Transition Show
Matthew Cloud, National Grid
Rohan Patel, Tesla

2:00 PM ET **Technical Talks and Q&A:** Electricity Generation and Load Profiles for EV Charging, and Related Impacts
Michael Kintner-Meyer, Pacific Northwest National Laboratory
Jeremy Michalek, Carnegie Mellon University
Yamen Nanne, Los Angeles Department of Water and Power

3:00 PM ET **Break**

3:10 PM ET **Technical Talks and Q&A:** Electricity Distribution System Loads for EV Charging
Deepak Divan, Georgia Institute of Technology
Eric Cutter, E3
Karen Glitman, Center for Sustainable Energy
Jason Hills, Los Angeles Department of Water and Power

4:15 PM ET **Panel Discussion:** Policy, Technical, and Market Options for Electricity System Needs
Sara Baldwin, Energy Innovation
Maria Bocanegra, Illinois Commerce Commission
Daniel Bowermaster, Electric Power Research Institute
Alan Jenn, University of California, Davis

5:15 PM ET Adjourn for the Day

DAY 4—OCTOBER 28, 2021
MEETING CONSUMER NEEDS

1:00 PM ET **Welcome**
Kate Whitefoot, Carnegie Mellon University, moderator

1:05 PM ET **Technical Talks and Q&A:** What Does the Future of EV Adoption Look Like from a Consumer Perspective?
Tyson Jominy, J.D. Power
Ahmed Abdulla, Carleton University
Shelley Francis, EV Noire

2:00 PM ET **Technical Talks and Q&A:** What Is Needed to Make EVs Available to All Buyers, Especially in the Used Market, and for Low-Income Drivers?
Gil Tal, University of California, Davis
Zhenhong Lin, Oak Ridge National Laboratory

3:00 PM ET **Break**

3:15 PM ET **Technical Talks and Q&A:** How Can EVs Meet the Transportation Needs of All Drivers, Including "Super-Users"?
Matthew Metz, Coltura
Nicholas Millar, EV Connect
Kelly Yearick, Forth

4:15 PM ET **Panel Discussion**: Policy, Technical, and Market Options to Remove Barriers to EVs Once Cost Parity Has Been Reached
Carla Bailo, Center for Automotive Research
Britta Gross, RMI
Sandra Wappelhorst, International Council on Clean Transportation
Yan (Joann) Zhou, Argonne National Laboratory

5:15 PM ET Workshop Adjourns

B

Biographies of Planning Committee Members

GARY MARCHANT, *Chair*, is a Regents Professor of Law and the director of the Center for Law, Science and Innovation at Arizona State University (ASU). His research interests include legal aspects of genomics and personalized medicine, the use of genetic information in environmental regulation, risk and the precautionary principle, and governance of emerging technologies such as nanotechnology, neuroscience, biotechnology, and artificial intelligence. He teaches courses in law; science and technology; genetics and the law; biotechnology: science, law, and policy; health technologies and innovation; privacy; big data and emerging technologies; and artificial intelligence: law and ethics. He was named a Regents Professor in 2011 and also is a professor in ASU's School of Life Sciences, a Distinguished Sustainability Scientist in ASU's Julie Ann Wrigley Global Institute of Sustainability, and is a Lincoln Professor of Emerging Technologies Law and Ethics with the Lincoln Center for Applied Ethics at ASU. Prior to joining ASU in 1999, Marchant was a partner at the Washington, DC, office of Kirkland & Ellis, where his practice focused on environmental and administrative law. During law school, he was the Editor-in-Chief of the *Harvard Journal of Law & Technology* and the editor of the *Harvard Environmental Law Review*, and was awarded the Fay Diploma (awarded to the top graduating student at Harvard Law School). Marchant frequently lectures about the intersection of law and science at national and international conferences. He has authored more than 150 articles and book chapters on various issues relating to emerging technologies. Among other activities, he has served on five previous National Academies of Sciences, Engineering, and Medicine committees (including the Committee on Assessment of Technologies and Approaches for Reducing the Fuel Consumption of Medium- and Heavy-Duty Vehicles, Phase 2, February 2013-present; the Committee on Assessment of Solid State Lighting, May 2011-February 2013; and the Committee on State Practices in Setting Mobile Source Emission Standards, 2004-2006), has been the principal investigator on several major grants, and has organized numerous academic conferences on law and science issues. He has a Ph.D. in genetics from the University of British Columbia.

CARLA BAILO is the president and chief executive officer (CEO) of the Center for Automotive Research (CAR). Bailo is a leader in engineering and vehicle program management with 35 years of experience in the automotive industry. In addition to her role at CAR, Bailo is the 2016-2018 vice president of automotive for SAE International, a global association of more than 138,000 engineers and related technical experts in the aerospace, automotive, and commercial vehicle industries. Prior to joining CAR, Bailo was the assistant vice president for mobility research and business development at The Ohio State University. In that capacity, she assisted the university in accelerating sustainable mobility and transportation innovation, while integrating related research and education across Ohio State's academic units. She also helped coordinate Ohio State's involvement as the primary research partner for Smart Columbus, a $140 million program to transform central Ohio into a premier transportation innovation region. Bailo has 25 years of experience at Nissan North America, Inc., where in her most recent role at Nissan she served as the senior vice president of research and development (R&D). She was responsible for vehicle engineering and development operations in Michigan, Arizona, Mexico, and Brazil, managing a $500 million budget and 2,500 employees. In this role, she improved the efficiency of Nissan's R&D functions. Bailo has an M.S. in mechanical engineering from the University of Michigan and a B.S. in mechanical engineering from Kettering University.

NADY BOULES is the president of NB Motors, LLC. He began his engineering and management consultancy practice with a focus on electrified, connected, and autonomous vehicles in September 2013 after 45 years of engineering experience (32 years with General Motors and Delphi), including 14 years as the director of R&D and innovation. In his most recent capacity as the director of GM's Research & Development's Electrical & Control Systems Research Lab, Boules was responsible for the development of advanced electrical systems and components for electrified, connected, and automated vehicles to enhance vehicle safety, comfort, and efficiency. He led all R&D activities in the areas of electronics and control software globally and coordinated research in this area, internally with sister research laboratories and engineering customers; and externally with collaborators at universities, national laboratories, suppliers, and other automotive original equipment manufacturers (OEMs) (United States Council for Automotive Research). Boules first joined General Motors Research Laboratories in 1982. He held several positions leading and managing research activities in automotive mechatronic and electric drive systems. In September 1999, he was named the director of R&D for Delphi Steering Systems in Saginaw, Michigan. His responsibilities expanded to cover brakes and suspension when he was named the director of the dynamics innovation center in 2002. In 2005, he was named the director, dynamics innovation center and materials engineering and his responsibilities expanded to materials for the Energy and Chassis division. From May 2006 until returning to GM, he held the position of director, innovation and technology leadership. Boules received his doctorate of engineering degree in 1978 from the Technical University of Braunschweig, Germany. He is the author of numerous patents and technical and invited papers and has received several awards from GM in recognition of his accomplishments, including the John Campbell Award (for scientific accomplishments), the Charles McCuen Award (recognizing contributions to the business success of GM), the Extraordinary Accomplishment Award, and the President's Council Honors Award. Boules has been a fellow of the Institute of Electrical and Electronics Engineers (IEEE) since 1991 and was named a life fellow in January 2015. He is also the recipient of the 2011 IEEE Nikola Tesla Award. He is a member of the Industry Applications Society and a past member of its executive board. He was also a member of the board of directors of the Intelligent Transportation Society of America (ITS-A) and a member of the executive board of several university consortia, and currently serves on the Committee on Review of the U.S. DRIVE Research Program of the National Academies.

ZHENHONG LIN is a senior R&D staff member at the National Transportation Research Center of Oak Ridge National Laboratory and a joint associate professor of industrial and systems engineering at the University of Tennessee. As the principal investigator and manager of the Transportation Energy Evolution Modeling program, he is interested in analyzing technologies, strategies, and policies that can transform the transportation energy system for societal objectives. Lin received the 2017 Vehicle Technologies Office Distinguished Achievement Award for his contributions to a joint study on Connected and Automated Vehicle energy impact. He also received the 2011 U.S. Department of Energy (DOE) Vehicle Technologies Office R&D Award for his work on modeling consumer choices of advanced powertrain technologies. Having authored more than 30 peer-reviewed articles, Lin has studied electric vehicle (EV) market acceptance, range optimization for battery and fuel cell EVs, on-road fuel economy, alternative fuel infrastructure optimization, diesel engine injection control, and fuel cell powertrain simulation. Lin is a member of the Alternative Transportation Fuels and Technologies' Committee of the Transportation Research Board and a senior member of IEEE. He received his Ph.D. in transportation engineering in 2008 and his M.S. in transportation technology and policy from the University of California (UC), Davis. Before that, Lin obtained his B.E. and M.S. in automotive engineering from Tsinghua University in Beijing.

JOSHUA LINN is an associate professor at the University of Maryland and a senior fellow at Resources for the Future (RFF). Linn's research centers on the effect of environmental regulation and market incentives on technology, with particular focus on the electricity sector and markets for new vehicles. His work on the electricity sector has compared the effectiveness of cap and trade and alternative policy instruments in promoting new technology, including renewable electricity technologies. Studies on new vehicles markets investigate the effect of Corporate Average Fuel Economy (CAFE) standards and fuel prices on new vehicle characteristics, technology, consumer well-being, and manufacturer profits. He has published in leading general interest and field journals in environmental, energy, and health economics. Linn, who joined the University of Maryland in 2018, joined RFF in March 2010, was an assistant professor in the Department of Economics at the University of Illinois at Chicago and a research scientist at the Massachusetts Institute of Technology (MIT). Linn holds a Ph.D. in economics from MIT and a B.A. in astronomy and physics from Yale University.

ANNA STEFANOPOULOU is the William Clay Ford Professor of Manufacturing at the University of Michigan. She has been on the faculty of the Department of Mechanical Engineering since 2000. She obtained her diploma (1991, National Technical University of Athens, Greece) in naval architecture and marine engineering and her Ph.D. (1996, University of Michigan) in electrical engineering and computer science. She served as the director of the Automotive Research Center, a multi-university U.S. Army Center of Excellence in Modeling and Simulation of Ground Vehicles (2009-2018). She was an assistant professor (1998-2000) at UC Santa Barbara, and a technical specialist (1996-1997) at Ford Motor Company, where she developed and implemented multivariable controllers for advanced engines and powertrains. She has been recognized as a fellow of three different societies; the American Society of Mechanical Engineers (ASME) (2008), IEEE (2009), and SAE (2018). She is an elected member of the executive committee of the ASME Dynamics Systems and Control Division and the board of governors of the IEEE Control Systems Society. She is the founding chair of the ASME Dynamic Systems and Control Division (DSCD) Energy Systems Technical Committee and a member of the National Academies' Committee on Assessment of Technologies for Improving Fuel Economy of Light-Duty Vehicles Phase 3. She is a recipient of the 2018 Rackham Distinguished Graduate Mentor Award, the 2017 IEEE Control System Technology award, the 2012 College of Engineering Research Award, the 2009 ASME Gustus L. Larson Memorial Award, a 2008 University of Michigan Faculty Recognition award, the 2005 Outstanding Young Investigator by the ASME DSCD, a 2005 Henry Russel award, a 2002 Ralph Teetor SAE educational award, a 1997 National Science Foundation CAREER award and selected as one of the 2002 world's most promising innovators from the MIT Technology Review. She has co-authored a book, 20 U.S. patents, and more than 250 publications (5 of which have received awards) on estimation and control of internal combustion engines and electrochemical processes such as fuel cells and batteries.

DEIDRE STRAND is the chief scientific officer at Wildcat Discovery Technologies. Strand has more than 25 years of experience in materials research, development, and commercialization, primarily in the areas of energy storage (lithium ion batteries) and electronic applications. Prior to joining Wildcat in 2012, she served as a research fellow at Dow Chemical, where she was the technical lead in Dow Energy Materials, as well as the principal investigator on external research programs with universities and national labs on battery materials. Strand also has extensive experience in patent analysis and technical due diligence of new technologies. She completed her Ph.D. in analytical chemistry at the University of Wisconsin–Madison, under the supervision of Professor John Schrag. Her Ph.D. research focused on rheology and birefringence of polymeric solutions. Strand also holds an M.S. in chemistry from the California Institute of Technology and a B.S. in chemistry from North Dakota State University.

KATE WHITEFOOT is an associate professor of mechanical engineering and engineering and public policy at Carnegie Mellon University. She is a thrust leader of Technology Commercialization for the Next Manufacturing Center and a faculty affiliate at the Carnegie Mellon Scott Institute for Energy Innovation. Prior to her current position, she served as a senior program officer and the Robert A. Pritzker fellow at the National Academy of Engineering (NAE), where she directed the NAE's Manufacturing, Design, and Innovation program. Whitefoot's research bridges engineering design theory and analysis with that of economics to study the design and manufacture of energy efficient and low-carbon products and processes and their adoption in the marketplace. Her areas of expertise include vehicle fuel efficiency, consumer choice, design and adoption of green products, energy-efficient and productive manufacturing, and energy and environmental policies. Whitefoot has gained recognition nationally and internationally for her research and teaching. She served on the National Academies' Committee on the Review of the National Institute of Standards and Technology Engineering Laboratory. Her research is featured in *The Washington Post*, *Popular Mechanics*, *Bloomberg Business*, and *Business Insider*, and referenced in the 2017-2025 Corporate Average Fuel Economy rulemaking. She has worked with several companies in automotive, aerospace, and high-tech industries, and has been invited to present briefings at the White House, Capitol Hill, the U.S. Department of Commerce, and the U.S. Environmental Protection Agency. Whitefoot earned three degrees from the University of Michigan: a B.S. and an M.S. in mechanical engineering and a Ph.D. in design science—a multidisciplinary program where she concentrated in mechanical engineering and economics, completing course sequences and having an advisory committee across both disciplines.

C

Biographies of Workshop Speakers

DAY 1: ELECTRIC VEHICLES 101 AND TRANSPORTATION DECARBONIZATION

NADY BOULES is the president of NB Motors, LLC. He began his engineering and management consultancy practice with a focus on electrified, connected, and autonomous vehicles in September 2013 after 45 years of engineering experience (32 years with General Motors and Delphi), including 14 years as the director of R&D and innovation. In his most recent capacity as the director of GM's Research & Development's Electrical & Control Systems Research Lab, Boules was responsible for the development of advanced electrical systems and components for electrified, connected, and automated vehicles to enhance vehicle safety, comfort, and efficiency. He led all R&D activities in the areas of electronics and control software globally and coordinated research in this area, internally with sister research laboratories and engineering customers; and externally with collaborators at universities, national laboratories, suppliers, and other automotive original equipment manufacturers (OEMs) (United States Council for Automotive Research). Boules first joined General Motors Research Laboratories in 1982. He held several positions leading and managing research activities in automotive mechatronic and electric drive systems. In September 1999, he was named the director of R&D for Delphi Steering Systems in Saginaw, Michigan. His responsibilities expanded to cover brakes and suspension when he was named the director of the dynamics innovation center in 2002. In 2005, he was named the director, dynamics innovation center and materials engineering and his responsibilities expanded to materials for the Energy and Chassis division. From May 2006 until returning to GM, he held the position of director, innovation and technology leadership. Boules received his doctorate of engineering degree in 1978 from the Technical University of Braunschweig, Germany. He is the author of numerous patents and technical and invited papers and has received several awards from GM in recognition of his accomplishments, including the John Campbell Award (for scientific accomplishments), the Charles McCuen Award (recognizing contributions to the business success of GM), the Extraordinary Accomplishment Award, and the President's Council Honors Award. Boules has been a fellow of the Institute of Electrical and Electronics Engineers (IEEE) since 1991 and was named a life fellow in January 2015. He is also the recipient of the 2011 IEEE Nikola Tesla Award. He is a member of the Industry Applications Society and a past member of its executive board. He was also a member of the board of directors of the Intelligent Transportation Society of America (ITS-A) and a member of the executive board of several university consortia, and currently serves on the Committee on Review of the U.S. DRIVE Research Program of the National Academies.

STEVEN CLIFF, as acting administrator of the National Highway Traffic Safety Administration (NHTSA), oversees the nation's vehicle safety agency that sets vehicle safety standards, identifies safety defects and manages recalls, and educates Americans to help them drive, ride, and walk safely. Most recently, he served as the deputy executive officer at the California Air Resources Board, an organization he first joined in 2008 as an air pollution specialist. Since then, he held a variety of positions at the California Air Resources Board (CARB), eventually overseeing its climate program. Cliff's most recent work as deputy executive officer at CARB included program oversight of regulations for passenger vehicle emissions, medium- and heavy-duty engine emissions, implementation of vehicle and engine emissions and on-board diagnostics certification, transportation land-use planning and analysis, and incentive and investment programs for reducing emissions. Cliff played an active role at UC Davis, for nearly two decades, joining the school's Applied Sciences Department in 2001 as a research professor and later becoming affiliated with the school's Air Quality Research Center. Cliff received a bachelor's degree and doctorate in chemistry from UC San Diego and completed a postdoc on atmospheric sciences at UC Davis, Department of Land, Air and Water Resources.

DAVE COOKE is a senior vehicles analyst in the Clean Transportation Program, specializing in both light- and heavy-duty fuel economy. He conducts research on fuel efficiency technologies and the implications for oil consumption and greenhouse gas emissions across the transportation sector. Before joining the Union of Concerned Scientists (UCS), Cooke was a Mirzayan Science and Technology Policy Fellow and associate program officer with the National Academies, where his work with the Board on Energy and Environmental Systems focused on automotive technologies, including peer-reviewed consensus studies on the development of advanced technology vehicles by 2050 and the barriers facing electric vehicle (EV) deployment. Cooke received his Ph.D. in condensed matter physics in 2010 from the University of California (UC), Berkeley, where his dissertation focused on the fundamental science behind modern hard drive technology, exploring the electronic and magnetic properties of these novel material systems through microcalorimetry. Prior to this, he received his B.S. in physics from Harvey Mudd College in 2002 and his M.S. in physics from UC San Diego in 2004.

ALEXANDER EDWARDS, in addition to the overall course and direction of Strategic Vision, is a key architect of tools and software, which measures decision-making, human behavior, attitudes, and perceptions. He directs operations and provides support for advanced clinical qualitative and quantitative studies. This includes large quantitative studies like the Mobility Experience Study and parts such as the New Vehicle Experience Study (NVES). Launched in 1994, NVES is the largest new vehicle owner study in the United States with an average of 250,000 new vehicle owners participating on an annual basis. In addition to understanding the consumer's experience with the vehicle, NVES also covers who the customer is (from demographics to personal priorities and values) to how and why they considered, purchased and rejected the vehicles that interested them.

DENISE GRAY is president of LG Energy Solution Michigan Inc. Tech Center, the North American subsidiary of LG Chem (LGC), South Korea. In this position, her team provides battery solutions for automotive and non-automotive applications. Gray holds several board or director positions: LG Energy Solution Michigan Inc., Tenneco Automotive, Canadian National Railway, Board on Energy and Environmental Systems, and Original Equipment Supplier Association. Recently, Gray was appointed to the U.S. Department of Energy, Secretary of Energy Advisory Board. Prior to her current position at LG, Gray held automotive leadership positions with AVL List in Graz, Austria and Atieva in Redwood City, California. The vast majority of her professional career (30 years) was with General Motors, where she spearheaded efforts in vehicle electrical and powertrain systems controls and software, including battery systems. Gray holds a B.S.E.E. from Kettering University and an M.S.E.S. from Rensselaer Polytechnic Institute.

JOSHUA LINN is an associate professor at the University of Maryland and a senior fellow at Resources for the Future (RFF). Linn's research centers on the effect of environmental regulation and market incentives on technology, with particular focus on the electricity sector and markets for new vehicles. His work on the electricity sector has compared the effectiveness of cap and trade and alternative policy instruments in promoting new technology, including renewable electricity technologies. Studies on new vehicles markets investigate the effect of Corporate Average Fuel Economy (CAFE) standards and fuel prices on new vehicle characteristics, technology, consumer well-being, and manufacturer profits. He has published in leading general interest and field journals in environmental, energy, and health economics. Linn, who joined the University of Maryland in 2018, joined RFF in March 2010, was an assistant professor in the Department of Economics at the University of Illinois at Chicago and a research scientist at the Massachusetts Institute of Technology (MIT). Linn holds a Ph.D. in economics from MIT and a B.A. in astronomy and physics from Yale University.

GARY MARCHANT is a Regents Professor of Law and the director of the Center for Law, Science and Innovation at Arizona State University (ASU). His research interests include legal aspects of genomics and personalized medicine, the use of genetic information in environmental regulation, risk and the precautionary principle, and governance of emerging technologies such as nanotechnology, neuroscience, biotechnology, and artificial intelligence. He teaches courses in law; science and technology; genetics and the law; biotechnology: science, law, and policy; health technologies and innovation; privacy; big data and emerging technologies; and artificial intelligence: law and ethics. He was named a Regents Professor in 2011 and also is a professor in ASU's School of Life Sciences, a Distinguished Sustainability Scientist in ASU's Julie Ann Wrigley Global Institute of Sustainability, and is a Lincoln Professor of Emerging Technologies Law and Ethics with the Lincoln Center for Applied Ethics at ASU. Prior to joining ASU in 1999, Marchant was a partner at the Washington, DC, office of Kirkland & Ellis, where his practice focused on environmental and administrative law. During law school, he was the Editor-in-Chief of the *Harvard Journal of Law & Technology* and the editor of the *Harvard Environmental Law Review*, and was awarded the Fay Diploma (awarded to the top graduating student at Harvard Law School). Marchant frequently lectures about the intersection of law and science at national and international conferences. He has authored more than 150 articles and book chapters on various issues relating to emerging technologies. Among other activities, he has served on five previous National Academies of Sciences, Engineering, and Medicine committees (including the Committee on Assessment of Technologies and Approaches for Reducing the Fuel Consumption of Medium- and Heavy-Duty Vehicles, Phase 2, February 2013-present; the Committee on Assessment of Solid State Lighting, May 2011-February 2013; and the Committee on State Practices in Setting Mobile Source Emission Standards, 2004-2006), has been the principal investigator on several major grants, and has organized numerous academic conferences on law and science issues. He has a Ph.D. in genetics from the University of British Columbia.

PATTY MONAHAN was appointed by Governor Gavin Newsom in April 2019 to be one of the five commissioners on the California Energy Commission. She serves in the Energy Commission's Science/Engineering position and is the lead commissioner on transportation. She has dedicated her career toward advancing clean energy technologies and cutting harmful pollution, with a focus on equity. Before joining the Energy Commission, she served as the transportation program director at the Energy Foundation, where she guided campaigns across the United States that advanced electric transportation and more efficient vehicles. She worked at UCS in several roles, including director of the California office and deputy director of clean vehicles. She also worked at the U.S. Environmental Protection Agency, Lawrence Berkeley National Laboratory, and several energy-related consulting firms. Monahan received a B.S. from UC Berkeley and holds a master's degree in energy analysis and policy from the University of Wisconsin–Madison.

RACHAEL NEALER is the deputy director for transportation technology and policy at the White House Council on Environmental Quality, where she advises and coordinates the execution of the Biden-Harris administration's transportation priorities. She also is the chair of the Transportation Research Board's (TRB's) Alternative Fuels and Technologies Committee. Throughout her career, she has focused on researching transportation as a system of systems and developing strategy around how to decarbonize transportation through technology development in concert with supporting policies. Nealer has previously held various roles at the U.S. Department of Energy, most recently, the senior advisor to the three transportation offices: Bioenergy, Hydrogen and Fuel Cell, and Vehicle Technologies Offices. She has also worked in the non-profit sector at UCS researching the environmental impacts of EVs compared to gasoline vehicles over their life. Prior to UCS she worked at the Environmental Protection Agency in the Renewable Fuels Standard office and she received her joint Ph.D. in civil and environmental engineering and engineering and public policy from Carnegie Mellon University where she specialized in life-cycle environmental impacts of transportation. As a former chief of staff, a personal passion of hers is ensuring the federal government is recruiting and retaining high quality talent.

DANIEL SPERLING is the Distinguished Blue Planet Prize Professor of Civil Engineering and Environmental Science and Policy and founding director of the Institute of Transportation Studies at UC Davis (ITS-Davis). He has held the transportation seat on the California Air Resources Board since 2007 and served as chair of the TRB of the National Academies in 2015-2016. Among his many prizes are the 2018 Roy W. Crum award from TRB, its highest research award; and the 2013 Blue Planet Prize from the Asahi Glass Foundation Prize for being "a pioneer in opening up new fields of study to create more efficient, low-carbon, and environmentally beneficial transportation systems." He served twice as lead author for the Intergovernmental Panel on Climate Change (sharing the 2007 Nobel Peace Prize), testified eight times to the Congress, and has authored or co-authored more than 250 technical papers and 13 books. Perhaps his most memorable accomplishment will be 6 minutes on *The Daily Show* with Jon Stewart.

AK SROUJI is the chief technology officer at Romeo Power, an energy technology and manufacturing company, building the highest energy density battery packs for EVs. He directs all technical and technological developments, including IP strategy, roadmaps, and product development, internally and in collaboration with strategic partners. Previously, Srouji was developing advanced physico-chemical battery models at Robert Bosch's Research & Technology Center in Silicon Valley. He also worked on novel battery component development, spanning from next-gen lithium ion to all solid-state batteries. Throughout his energy storage technology career he has successfully brought to market multiple programs funded by the U.S. Department of Energy (DOE). He holds an M.S. in mechanical engineering and a Ph.D. in energy and mineral engineering with a focus on electrochemistry from The Pennsylvania State University.

DEIDRE STRAND is the chief scientific officer at Wildcat Discovery Technologies. Strand has more than 25 years of experience in materials research, development, and commercialization, primarily in the areas of energy storage (lithium ion batteries) and electronic applications. Prior to joining Wildcat in 2012, she served as a research fellow at Dow Chemical, where she was the technical lead in Dow Energy Materials, as well as the principal investigator on external research programs with universities and national labs on battery materials. Strand also has extensive experience in patent analysis and technical due diligence of new technologies. She completed her Ph.D. in analytical chemistry at the University of Wisconsin–Madison, under the supervision of Professor John Schrag. Her Ph.D. research focused on rheology and birefringence of polymeric solutions. Strand also holds an M.S. in chemistry from the California Institute of Technology and a B.S. in chemistry from North Dakota State University.

KATE WHITEFOOT is an associate professor of mechanical engineering and engineering and public policy at Carnegie Mellon University. She is a thrust leader of Technology Commercialization for the Next Manufacturing Center and a faculty affiliate at the Carnegie Mellon Scott Institute for Energy Innovation. Prior to her current position, she served as a senior program officer and the Robert A. Pritzker fellow at the National Academy of Engineering (NAE), where she directed the NAE's Manufacturing, Design, and Innovation program. Whitefoot's research bridges engineering design theory and analysis with that of economics to study the design and manufacture of energy efficient and low-carbon products and processes and their adoption in the marketplace. Her areas of expertise include vehicle fuel efficiency, consumer choice, design and adoption of green products, energy-efficient and productive manufacturing, and energy and environmental policies. Whitefoot has gained recognition nationally and internationally for her research and teaching. She served on the National Academies' Committee on the Review of the National Institute of Standards and Technology Engineering Laboratory. Her research is featured in *The Washington Post*, *Popular Mechanics*, *Bloomberg Business*, and *Business Insider*, and referenced in the

2017-2025 Corporate Average Fuel Economy rulemaking. She has worked with several companies in automotive, aerospace, and high-tech industries, and has been invited to present briefings at the White House, Capitol Hill, the U.S. Department of Commerce, and the U.S. Environmental Protection Agency. Whitefoot earned three degrees from the University of Michigan: a B.S. and an M.S. in mechanical engineering and a Ph.D. in design science—a multidisciplinary program where she concentrated in mechanical engineering and economics, completing course sequences and having an advisory committee across both disciplines.

DAY 2: VEHICLE PRODUCTION AND LIFE CYCLE

RENATA ARSENAULT is Ford Motor Company's technical expert for advanced battery recycling. She works at Ford Ion Park, Ford's new Battery Center of Excellence, prior to which she was in Ford's Research & Advanced Engineering organization. Arsenault has been interfacing closely with battery suppliers, recyclers and other industry members since 2006. She is involved in numerous facets of Sustainability at Ford, representing the battery group on battery recycling, second use and related topics (life-cycle analysis, strategic materials, supply chain circularity etc.). Arsenault serves as PM for numerous United States Advanced Battery Consortium programs, is president-elect of NAATBatt, co-chair of NAATBatt's Battery Recycling and NA Manufacturing committees, and manages projects with multiple universities as part of Ford's University Research program. She has been a speaker and panelist on a long list of global battery conferences and has numerous publications and patents. Arsenault graduated from Queen's University in Ontario, Canada in 1988 with a B.S. in chemical engineering. She spent 1989-1998 working in specialty chemicals for Diversey/Diversey-Lever where she led process development and scale-up of specialty products in North America.

FAN DAI is the executive director of the California-China Climate Institute at UC Berkeley. Dai has played a significant role leading California's collaboration with China on climate, energy, and environment. She was appointed by Governor Edmund G. Brown, Jr., as the special advisor on China. Under Brown, Dai chaired the state's China Interagency Working Group, and acted as the state's liaison on its critical economic and environmental initiatives on China. Previously, she served as the senior advisor at California Environmental Protection Agency and California Governor's Office of Business and Economic Development, advising on the state's international policy and global climate partnership. In 2017, she organized Governor Brown's trip, which resulted in a successful meeting with President Xi Jinping and the commitment to establish the California-China Climate Institute.

WARREN DAY is a research geologist with the U.S. Geological Survey (USGS) Mineral Resources Program and is based in Golden, Colorado. Day is currently the science coordinator for the USGS Earth Mapping Resources Initiative (commonly known as Earth MRI). Earth MRI is a partnership with the State Geological Surveys, other federal agencies, and private industry to acquire modern framework geologic, geophysical, and topographic data for areas across the nation that may host critical mineral resources. He received his undergraduate degree in Earth and planetary sciences from Washington University, St. Louis, Missouri and doctorate in geology and geophysics from the University of Minnesota. Over the past 38 years, Day has conducted research on regional framework geology, the evolution of ore-forming systems, and associated mineral resource studies as well as having held several science management positions with the USGS.

KRISTIN DZICZEK is the senior vice president of research at the Center for Automotive Research (CAR). Dziczek has more than 25 years of experience as a researcher and policy analyst. She is globally recognized as an expert on automotive competitiveness, trade, industrial policy, labor, and talent issues. Dziczek leads CAR's Business and Technology groups, with a research portfolio focused on developing a better understanding of the connections between the automotive industry, technology, the economy, society, and public policy. Before joining CAR, she served as the associate director of the Michigan Manufacturing Technology Center and has worked for the U.S. Congress, United Auto Workers (UAW) International Union, and General Motors Corporation. Dziczek earned her B.A. in economics, M.P.P. in public policy, and M.S. in industrial and operations engineering, all from the University of Michigan.

LINDA GAINES is the chief scientist of the new ReCell Center for Advanced Battery Recycling and a Systems Analyst in the Energy Systems Division at Argonne National Laboratory. She holds a B.A. in chemistry and physics from Harvard University, and a Ph.D. in physics from Columbia University. Her primary interest is problem solving, applied to efficient use of resources. She began her career at Argonne by writing a series of handbooks of energy and material flows in energy-intensive industries that provided background for studies of technical and institutional issues involved in producing and recycling advanced-design automobiles, trucks, and trains, and batteries. Her most recent work has involved studying ways to reduce impacts from transport by recycling of lithium-ion batteries. She is an editor of the journal *Sustainable Materials and Technologies* and on the editorial board of *Scientific Reports* (a *Nature* journal).

JOHN D. GRAHAM is a professor of risk analysis at the Paul H. O'Neill School of Public and Environmental Affairs, Indiana University. He formerly served as a senior Office of Management and Budge official in the George W. Bush administration. He published a book titled *The Global Rise of the Modern Plug-In Electric Vehicle: Public Policy, Innovation, and Strategy* (2021).

DAVID HOWELL is the acting Vehicles Technologies Office (VTO) director in the Office of Energy Efficiency and Renewable Energy with responsibility for the department's $400 million advanced vehicle R&D portfolio. Howell has more than 35 years of experience planning and executing complex, multi-disciplined R&D activities that includes hybrid and EV R&D, advanced battery research and manufacturing, and advanced structural materials research. He serves as the chair of the Federal Consortium for Advanced Batteries, which brings together federal agencies to accelerate the development of a resilient, robust lithium-ion battery supply chain. The consortium successfully developed the recently released National Blueprint for Lithium Batteries will help guide investments to develop a resilient lithium-battery manufacturing value chain. Howell served as DOE's program manager for battery R&D from 2003-2017. He continues to serve as DOE's representative at the United States Advanced Battery Consortium Management Committee, the executive committee for the Battery500 Research Consortium, and various international and inter-government forums. Prior to DOE, Howell was on the research staff of the Oak Ridge National Laboratory and served on active duty at the Wright Patterson Air Force Base as the program manager for Advanced Materials for Space Structures at the Air Force Materials Laboratory. He received a B.S. in aerospace engineering from the University of Tennessee, Knoxville.

DAVID KLANECKY has spent most of his career in senior operational, research and development, commercial, and strategic leadership roles. From 2013 to 2021, he served in increasingly senior management roles within Albemarle Corporation, the world's leading lithium producer, including as the vice president of strategy and corporate development and most recently serving as the vice president—Lithium Operations—APAC/EU, with global responsibility for Albemarle's manufacturing/operations, process technology and product management within the global lithium business. Klanecky also served as interim chief executive officer (CEO) of the MARBL joint venture between Albemarle and Mineral Resources Ltd, which includes the Kemerton and Wodgina assets in Australia. Before joining Albemarle, he had an impressive 20-year career with The Dow Chemical Company that spanned the globe. In his last role with Dow, Klanecky launched the Dow Energy Materials Business, focused on lithium-ion battery materials offerings to cell manufacturers and Auto OEMs, where he served as the global business director for this business unit prior to joining Albemarle in 2013. Klanecky has a chemical engineering degree from the University of Nebraska and an Executive M.B.A. from Arizona State/Thunderbird School of Global Management. He joined Piedmont Lithium in April 2021 and is based in the company's Belmont, North Carolina, headquarters.

ZOE LIPMAN is the director of Manufacturing and Advanced Transportation at the BlueGreen Alliance (BGA) where she leads BGA's policy and advocacy on clean technology manufacturing, advanced vehicles and transportation, and related economic topics. Prior to joining BGA, Lipman led the National Wildlife Federation's program on fuel economy and advanced and EVs, and previously headed the organization's Midwest climate policy program. She has written and advocated on advanced vehicle supply chains, manufacturing and jobs policy, transportation and utility sector innovation, and other opportunities to speed adoption of climate and energy solutions while rebuilding America's economy, manufacturing communities, jobs and competitiveness. Lipman has also served on utility, transportation and climate policy forums at a state, regional and federal level and works closely with labor, environmental, business, and government stakeholders. She holds a B.A. from Yale University and an M.P.A. from Harvard University's John F. Kennedy School of Government.

BRAD MARKELL serves as the executive director of both the AFL-CIO Industrial Union Council and the Working for America Institute. He also leads the federation's climate and energy policy work, and is a cleared trade advisor for the AFL-CIO. Prior to joining the AFL-CIO, Markell worked for the UAW, where he represented the union during the negotiations for the 2009 and 2012 tailpipe emissions standards, helped develop and pass the legislation establishing DOE's Advanced Technology Vehicles Manufacturing program, and participated in several rounds of national bargaining in the automobile, aerospace and heavy-truck industries. Markell's board and committee service on behalf of the labor movement has included the Biden-Harris transition team, the Labor and Employment Relations Association, the National Academy of Engineering, the Export-Import Bank Advisory Council, the Carbon Capture Coalition, the Michigan Climate Action Council, and the International Labor Organization's Expert Committee on Just Transition. He has degrees from the University of Michigan and Wayne State University. He joined the UAW in 1976 and is a member of Local 14 in Toledo, Ohio.

CELINA MIKOLAJCZAK, the vice president of manufacturing engineering at QuantumScape, leads teams spanning the battery development process—from materials sourcing and cell engineering to automation equipment design and environmental health and safety—and is responsible for bringing QuantumScape solid-state technology into mass production. Mikolajczak has held several leadership roles in her more than 20 years of working in Li-ion batteries. Prior to joining QuantumScape, she was the vice president of engineering and battery technology at Panasonic Energy of North America, which produces Li-ion cells for Tesla at the Gigafactory in Sparks, Nevada. Mikolajczak worked at Uber as the director of battery engineering, leading the design of battery packs for micro-mobility and urban air mobility applications. Before that, she led Cell Quality and Materials Engineering at Tesla as the company launched its Model S, Model X, Model 3, Powerwall, and Powerpack products. Mikolajczak has a particular passion for battery safety; she frequently publishes and speaks on the topic and has served on various Li-ion battery safety standards committees. Earlier in her career, she helped found battery consulting group Exponent and developed battery failure analysis techniques and safety standards related to consumer electronics devices.

ANNA STEFANOPOULOU is the William Clay Ford Professor of Manufacturing at the University of Michigan. She has been on the faculty of the Department of Mechanical Engineering since 2000. She obtained her diploma (1991, National Technical University of Athens, Greece) in naval architecture and marine engineering and her Ph.D. (1996, University of Michigan) in electrical engineering and computer science. She served as the director of the Automotive Research Center, a multi-university U.S. Army Center of Excellence in Modeling and Simulation of Ground Vehicles (2009-2018). She was an assistant professor (1998-2000) at UC Santa Barbara, and a technical specialist (1996-1997) at Ford Motor Company, where she developed and implemented multivariable controllers for advanced engines and powertrains. She has been recognized as a fellow of three different societies; the American Society of Mechanical Engineers (ASME) (2008), IEEE (2009), and SAE (2018). She is an elected member of the executive committee of the ASME Dynamics Systems and Control Division and the board of governors of the IEEE Control Systems Society. She is the founding chair of the ASME Dynamic Systems and Control Division (DSCD) Energy Systems Technical Committee and a member of the National Academies' Committee on Assessment of Technologies for Improving Fuel Economy of Light-Duty Vehicles—Phase 3. She is a recipient of the 2018 Rackham Distinguished Graduate Mentor Award, the 2017 IEEE Control System Technology award, the 2012 College of Engineering Research Award, the 2009 ASME Gustus L. Larson Memorial Award, a 2008 University of Michigan Faculty Recognition award, the 2005 Outstanding Young Investigator by the ASME DSCD, a 2005 Henry Russel award, a 2002 Ralph Teetor SAE educational award, a 1997 National Science Foundation CAREER award and selected as one of the 2002 world's most promising innovators from the MIT Technology Review. She has co-authored a book, 20 U.S. patents, and more than 250 publications (5 of which have received awards) on estimation and control of internal combustion engines and electrochemical processes such as fuel cells and batteries.

DAY 3: CHALLENGES AND SOLUTIONS FOR THE ELECTRIC SYSTEM

SARA BALDWIN is Energy Innovation's director of electrification policy, where she leads the firm's electrification policy practice area to advance economy-wide decarbonization and provide guidance to policymakers and regulators on viable pathways to advance the electrification of buildings, transportation, and industry. Baldwin is the lead author of *Accelerating Clean, Electrified Transportation by 2035: Policy Priorities* and contributing author to *Increasing Electric Vehicle Charging Access at Multi-Unit Dwellings: Workshop Summary Report*, among others. From 2014-2020, Baldwin served as the vice president of Regulatory for the Interstate Renewable Energy Council (IREC), overseeing IREC's engagement in regulatory proceedings in dozens of states and led the development of several landmark publications, including *A Playbook for Modernizing the Distribution*. She also served on the technical team for the DOE SunShot Solar Market Pathways Initiative and the NREL's Distributed Generation Interconnection Collaborative Advisory Committee. From 2004-2014, Baldwin led clean energy policy efforts in Utah as a senior policy associate for Utah Clean Energy. She was named an "Innovator and Influencer" by Solar Power World (2017) and one of "Utah's Enlightened 50" by the Community Foundation of Utah (2012). Baldwin is a member of GridLab's advisory board, host of Energy Innovation's Electrify This! podcast, and former host of the Grid Geeks podcast. She earned her B.S. in environmental studies and a B.A. in Spanish from the University of Utah, graduating with honors.

MARIA BOCANEGRA is a commissioner with the Illinois Commerce Commission where she regulates public utilities, transportation, railroad, and telecommunications. She currently serves as the national chair for the EV State working group at National Association of Regulatory Utility Commissioners and holds advisory positions with DOE's Voice of Experience's Electrification Steering Committee, the National EV Charging Initiative's Advisory Council and the World Resources Institute's Electric School Bus Initiative. Bocanegra has been recognized by Fortnightly's Under 40, Negocios Now's 40 Under 40, Negocios Now's Who's Who in Hispanic Chicago, Negocios Now's Rising Star, and by Latino Leaders Magazine. In 2021, Negocios Now selected her as one of the 50 Most Influential Latinos in Chicagoland. She is only the second Latina to be appointed to the Illinois Commerce Commission.

DANIEL BOWERMASTER is the senior program manager for Electric Transportation at the Electric Power Research Institute (EPRI), an independent non-profit center for public interest energy and environmental research. For more than 30 years, EPRI's electric transportation program provides research critical to the development of technology, infrastructure, and analytics necessary to support the advancement of electric transportation. Bowermaster has worked on EVs for more than 12 years, including 2 years at PG&E prior to EPRI. He completed the Wharton-Lauder dual graduate degree program, earning an M.B.A. from the Wharton School of Business and an M.A. in international studies from the University of Pennsylvania. In addition, he holds a B.S. in mechanical engineering and a B.A. in international relations from UC Davis.

MATTHEW CLOUD received his B.Sc. in electrical engineering from Clarkson University in 2010 and an M.S.E.E. from the Worcester Polytechnic Institute in 2016. He is currently a lead engineer in the National Grid's Future of Electric organization developing the company's strategy to support and promote EV adoption. Previously he spent 3 years in distribution planning within National Grid after spending the prior 7 years working in various engineering roles at The United Illuminating Company in Orange, Connecticut. He has studied the adoption of EVs and their impacts to electric utility networks for the past 10 years and has co-authored three published papers on the modeling of EVs and the efficacy of time-of-use rates in reducing grid impacts from EV charging.

ERIC CUTTER is a senior director at Energy + Environmental Economics (E3) leading our distributed energy resources practice area, which focuses on enabling energy storage, EVs, and flexible loads to respond to grid needs. Cutter's work demonstrates how flexible and dispatchable distributed energy resources supports higher penetrations of renewable generation and provides value to utilities and ratepayers. He is currently working with utilities in California, New York, and the Pacific Northwest to assess the impact of EVs on distribution system costs and to develop dynamic charging programs that better accommodate renewable generation under future high-renewables scenarios.

DEEPAK DIVAN is a professor, the John E. Pippin Chair, a Georgia Research Alliance Eminent Scholar, and the director of the Center for Distributed Energy at the Georgia Institute of Technology. His field of research is in the areas of power electronics, power systems, smart grids, electrification of transportation and distributed control of power systems. He works closely with utilities and industry and is actively involved in research, teaching, entrepreneurship and starting new ventures. Divan has started several companies, including Varentec in Santa Clara, California, where he served as the founder, president, and chief technology officer from 2011-2014, and as the chief scientist for several years after. He led the company as it developed its suite of innovative distributed real-time grid control technologies. Varentec was funded by leading green-tech Venture Capital firm Khosla Ventures and renowned investor Bill Gates. He has founded or seeded several new ventures including Soft Switching Technologies, Innovolt, Varentec and Smart Wires, which together have raised more than $160 million in venture funding. Divan is an elected member of the National Academy of Engineering, a member of the National Academies' Board on Energy and Environmental Systems, and the Committee on the Future Grid. He is a fellow of the IEEE, past president of the IEEE Power Electronics Society, is a recipient of the IEEE William E. Newell Field Medal and is International Steering Committee Chair of the IEEE Empower a Billion Lives global competition to crowdsource scalable energy access solutions. He has 40 years of academic and industrial experience, 75 issued and pending patents, and more than 400 reviewed publications. He received his B.Tech. from IIT Kanpur, and his M.S. and Ph.D. from the University of Calgary, Canada.

KAREN GLITMAN is the senior director of Distributed Energy Resources at the Center for Sustainable Energy (CSE). She has more than 35 years of policy advocacy and public service experience, having served as a state legislator, deputy secretary of transportation, director of the University of Vermont University Transportation Center, and director of Efficiency Vermont. Currently she leads CSE efforts to uncover new opportunities to support entities around the nation in meeting a shared goal of decarbonization.

JASON HILLS is the manager of distribution and electrification at the Los Angeles Department of Water and Power (LADWP). LADWP supports the growth in adoption of EVs and currently has more than 14,000 public, workplace, and fleet charging stations in its territory, achieving its 2022 goal of 10,000 EV chargers nearly 2 years ahead of schedule. In this role, Hills oversees the implementation of various electric transportation programs that include the installation of charging stations at LADWP properties and the development of customer-facing EV incentive programs. He also oversees LADWP's distribution engineering team and has previously managed substation programs aimed at supporting the reliability and growth of LADWP's Power System.

ALAN JENN is currently the assistant director of the Energy Futures Research Center, an assistant professional researcher at the Plug-in Hybrid and Electric Vehicle group of the Institute of Transportation Studies (ITS) at UC Davis, and an affiliate at Lawrence Berkeley National Laboratory. He graduated from Carnegie Mellon University with a Ph.D. in the engineering and public policy and has undergraduate degrees in molecular and cell biology, music, and energy and resources from UC Berkeley. Jenn's research is focused on plug-in EVs: integration with the electric grid, adoption of the technology, use in ride-hailing companies (such as Uber and Lyft), and its impact on transportation finance.

MICHAEL KINTNER-MEYER is a research engineer and systems analyst with more than 25 years in the area of energy/economy and multi-sector simulation and analysis for national and international clients. Educated in Germany with a Diplom Ingenieur degree from RWTH Aachen in mechanical engineering and a Ph.D. from the

University of Washington, Seattle, he works on European Union-funded Energy/Economic research and on the U.S. Energy Information Administration's National Energy Modeling System (NEMS). He is currently leading the mobility and electrification of transportation research area as well as Pacific Northwest National Laboratory's (PNNL's) energy policy research and institutional support for the Grid Modernization Initiative of the U.S. Department of Energy. He has led several major national assessments for energy storage, Hyperloop technology impacts on the U.S. Grid, EV grid impacts, and valuation of EVs for renewables integration. Currently, Kintner-Meyer leads for PNNL the research activities in mobility research focusing on transitioning the transportation sector to meet the nation's 21st-century mobility needs under sustainability objectives. This includes decarbonization strategies, large-scale traffic simulations of urban centers, connected vehicle controls research, as well as EV charging controls technology development. Kintner-Meyer holds nine patents on EV charge management, grid-friendly appliances, and energy storage controls technologies.

JEREMY J. MICHALEK is a professor of engineering and public policy, a professor of mechanical engineering, and the director of the Vehicle Electrification Group at Carnegie Mellon University. He has published widely on technical, economic, environmental, behavioral, and public policy aspects of EVs, ridesourcing, new mobility, transportation policy, and other topics. Michalek earned his B.S. from Carnegie Mellon University in 1999 and his M.S. and Ph.D. from the University of Michigan in 2001 and 2005. His research is featured frequently in news outlets, such as *The New York Times* and *National Public Radio*, and he engages policymakers at federal and state levels, including via policy briefings and public comments on proposed rulemaking. Michalek is currently serving on the National Academies' Committee on Current Methods for Life Cycle Analysis of Low-Carbon Transportation Fuels in the United States, and he serves on the Alternative Transportation Fuels and Technologies committee of the Transportation Research Board.

YAMEN NANNE is the manager of distribution system development at LADWP. He graduated from Cal Poly Pomona with a degree in electronics and computer engineering. His passion about EVs stems from his desire to play a role in improving air quality for all of California and to help LADWP customers learn about how they can save money by driving and EV. His motto is, Live Life... Drive Electric! As LADWP's Electric Transportation Program supervisor, he was responsible for leading the utility's effort to enable its customers and the City of Los Angeles to become the leader in Electric Transportation. During his tenure LADWP and the City of Los Angeles, he was able to increase EV charging infrastructure 11-fold from about 1,200 commercial charging stations in 2018 to 13,900 today and EV adoption increased from 35,000 EVs to about 73,000 EVs today.

CHRIS NELDER is the creator and host of the Energy Transition Show podcast, and a full-time traveler. Nelder has written about energy transition for more than 15 years. He is the author of two books on energy and investing, as well as more than 200 articles on energy in publications such as *Nature*, *Scientific American*, *Slate*, *The Atlantic*, *Quartz*, *Financial Times*, Greentech Media, and SmartPlanet. Prior to taking the podcast on the road full time, he headed the EV-grid Integration team for 5 years in the Mobility practice at RMI in Boulder, Colorado. He enjoys bantering with other energy geeks on Twitter at @chrisnelder.

ROHAN PATEL serves as the senior global director of public policy and business development at Tesla, where he oversees Tesla's regulatory, business development, and legislative efforts. Previously, he served as Special Assistant to the President and Senior Advisor for Climate and Energy Policy, where he worked on a range of power sector and transportation policies in addition to being President Obama's primary liaison with local elected officials. Patel also served as the associate director at the White House Council on Environmental Quality during the first term of the Obama administration. After the 2008 inauguration, he joined the Obama administration as the senior policy advisor to Secretary Vilsack at the U.S. Department of Agriculture. Prior to his appointment, he spent 2 years on the Obama Presidential Campaign in a number of leadership roles. He has also served on the staffs of Governor Chet Culver (D-IA), the Iowa Senate Democrats, and Senator Evan Bayh (D-IN). Patel is a graduate of Northwestern University and a native of Goshen, Indiana.

DAY 4: MEETING CONSUMER NEEDS

AHMED ABDULLA is an assistant professor in the Department of Mechanical and Aerospace Engineering at Carleton University. Abdulla investigates energy system design for deep decarbonization—focusing on the role of disruptive energy technologies that sit at a low level of technical readiness, including electro-fuel production, advanced nuclear power, and negative emissions technologies. Abdulla employs chemical process modeling, systems engineering, engineering economics, and quantitative risk and decision analysis in his research. He also pays special attention to integrating insights from public policy and behavioral science in his models to optimize the design and deployment of truly sustainable technologies—ones that are both techno-economically viable and socio-politically acceptable. Abdulla co-leads the APEX research group at Carleton. APEX—Alternative Pathways for the Energy Transition—comprises a group of highly interdisciplinary engineers devoted to accelerating the transition to a deeply decarbonized energy system to avert the worst consequences of climate change.

CARLA BAILO is the president and CEO of the Center for Automotive Research (CAR). Bailo is a leader in engineering and vehicle program management with 35 years of experience in the automotive industry. In addition to her role at CAR, Bailo is the 2016-2018 vice president of automotive for SAE International, a global association of more than 138,000 engineers and related technical experts in the aerospace, automotive, and commercial vehicle industries. Prior to joining CAR, Bailo was the assistant vice president for mobility research and business development at The Ohio State University. In that capacity, she assisted the university in accelerating sustainable mobility and transportation innovation, while integrating related research and education across Ohio State's academic units. She also helped coordinate Ohio State's involvement as the primary research partner for Smart Columbus, a $140 million program to transform central Ohio into a premier transportation innovation region. Bailo has 25 years of experience at Nissan North America, Inc., where in her most recent role at Nissan she served as the senior vice president of R&D. She was responsible for vehicle engineering and development operations in Michigan, Arizona, Mexico, and Brazil, managing a $500 million budget and 2,500 employees. In this role, she improved the efficiency of Nissan's R&D functions. Bailo has an M.S. in mechanical engineering from the University of Michigan and a B.S. in mechanical engineering from Kettering University.

SHELLEY FRANCIS, a former medical school faculty member, is a public health executive, entrepreneur and transportation disruptor. Francis is a co-founder and principal at EVNoire. EVNoire's work centers on shared, connected, electric and autonomous vehicle technology. EVNoire's work focuses on two pillars: E-Mobility Best Practices and E-Mobility Diversity, Equity and Inclusion. Francis and Terry Travis' vision led to the co-founding of the nation's largest network of diverse EV drivers and enthusiasts, EVHybridNoire, a 501c3 Nonprofit. EVHybridNoire, is a national, award winning, multicultural organization focused on increasing EV adoption and awareness in diverse communities. Francis is frequently sought out for her expertise in E-Mobility, the public health impacts of transportation emissions and environmental justice and is often featured in national journals and publications such as the *Baltimore Sun*, *Columbus Times-Dispatch*, *Energy News*, *Forbes*, and more. Francis was recently recognized as one of the top Women in EVs by Electric Drives; she was also featured in the Progressive Change Institute's report of the top 400 candidates recommended for the Biden-Harris administration and she was identified by movmi and Sandra Phillips as one of their 2020 Mobility Leaders. She currently serves as a national board director for the Electric Auto Association.

BRITTA GROSS is the managing director of RMI's mobility practice area, focused on the market-driven strategies and technologies required to accelerate toward carbon-free mobility solutions. Gross was formerly the director of advanced vehicle commercialization at General Motors, responsible for the energy and infrastructure strategies required to enable the commercialization of battery electric and hydrogen fuel cell EVs. Gross is also currently the president of Orland's Utility Commission's board of commissioners, Orlando's electric and water utility. She has an electrical engineering degree from LSU and studied language arts at the University of Wurzburg in Germany. She holds and has held numerous board seats, including the Alliance for Transportation Electrification, the North American Council for Freight Efficiency, MobilityData, Plug in America, the

Electric Drive Transportation Association and served as a governor appointee on both the Massachusetts Zero Emission Vehicle Commission and the Maryland Electric Vehicle and Infrastructure Commission. Gross has received numerous industry awards including the Forth Roadmap Career Achievement Award, the *Automotive News* "Electrifying 100," the GreenBiz "Verge 25" award, has testified to the Senate Committee on Energy and Natural Resources, and speaks regularly to national audiences on topics related to transportation electrification.

TYSON JOMINY is the vice president of data and analytics at J.D. Power. He has worked for J.D. Power since September 2009, advising OEMs on pricing and incentives for new and used vehicle sales. Prior to joining J.D. Power, Jominy worked for Ford Motor Company and Nissan North America, where he held several positions within finance and sales & marketing. Additionally, he has managed financial forecasting and planning in the consumer packaged goods industry and advised executives in corporate development and mergers and acquisitions in the energy sector. Jominy has a bachelor's degree in finance from Indiana University in Bloomington, Indiana, and an M.B.A. from the University of Chicago Booth Graduate School of Business.

ZHENHONG LIN is a senior R&D staff member at the National Transportation Research Center of Oak Ridge National Laboratory and a joint associate professor of industrial and systems engineering at the University of Tennessee. As the principal investigator and manager of the Transportation Energy Evolution Modeling program, he is interested in analyzing technologies, strategies, and policies that can transform the transportation energy system for societal objectives. Lin received the 2017 Vehicle Technologies Office Distinguished Achievement Award for his contributions to a joint study on Connected and Automated Vehicle energy impact. He also received the 2011 DOE Vehicle Technologies Office R&D Award for his work on modeling consumer choices of advanced powertrain technologies. Having authored more than 30 peer-reviewed articles, Lin has studied EV market acceptance, range optimization for battery and fuel cell EVs, on-road fuel economy, alternative fuel infrastructure optimization, diesel engine injection control, and fuel cell powertrain simulation. Lin is a member of the Alternative Transportation Fuels and Technologies' Committee of the Transportation Research Board and a senior member of IEEE. He received his Ph.D. in transportation engineering in 2008 and his M.S. in transportation technology and policy from UC Davis. Before that, Lin obtained his B.E. and M.S. in automotive engineering from Tsinghua University in Beijing.

MATTHEW METZ is the founder and co-executive director of Coltura. Before starting Coltura in 2014, he was a trial lawyer and non-profit leader. Metz has been at the forefront of the anti-gasoline movement since its inception. Along with his co-executive director Janelle London, he wrote the first op-eds in the United States calling for a phaseout of gasoline-powered vehicles, sponsored early gatherings of California and global leaders to discuss 100 percent vehicle electrification, and led coalitions to support the passage of gasoline vehicle phaseout legislation in California and Washington State. He writes about legal, political, and cultural strategies to transition off gasoline, most recently the gasoline superusers report. He was the lead author of recent law review articles about gas stations and state EV mandates. Metz has created and produced performance and conceptual art works aimed at changing cultural perceptions of gasoline. Prior to founding Coltura, he founded the Metz Law Group, Jaguar Forest Coffee Company, the Central Area Development Association and the Central Area Arts Council, and worked for an international development non-governmental organization in Mexico City. He is a graduate of the University of Chicago and UC Los Angeles Law School and is fluent in Spanish.

NICHOLAS MILLAR is a product manager at EV Connect, focusing on growing the availability of EV charging stations through software that enables great experiences for drivers and charging station managers alike. He has a background in mechanical engineering and has worked in many different applications of software for the energy transition, including energy efficiency, stationary battery storage, and EV charging.

GIL TAL holds a Ph.D. in transportation technology and policy from UC Davis, and an M.A. in geography and environmental policy and planning from the Hebrew University in Jerusalem. Between 2008 and 2010, Tal was a post-doctoral researcher with the Center for Global Metropolitan Studies and the UC Transportation Center at UC Berkeley. His work at UC Davis and UC Berkeley focuses on travel behavior and the implementation of travel behavior related policies. At the PH&EV center, Tal is leading projects on the future need for EV infrastructure, and the correlation between charging infrastructure, travel behavior and the demand for EVs. He is currently leading research on number of projects including a study on local planning and deployment of EV infrastructure, a study on geographic information system tools for infrastructure planning, a multi-state study of new plug-in vehicle buyers, and a study on the secondary market of plug-in vehicles in California.

SANDRA WAPPELHORST is a senior researcher on the International Council on Clean Transportation's (ICCT's) Europe team. Her current research focuses on financial and non-financial incentive systems for EVs and the electrification of shared ride services. Prior to joining the ICCT, Wappelhorst worked for various research organizations with a focus on urban development, the adoption of innovative mobility services, and user acceptance of EVs. She holds a diploma degree as an engineer in spatial planning (Dipl.-Ing.) from the Technical University of Dortmund (Germany) and a doctor's degree (Dr.-Ing.) from the University of the Federal Armed Forces Munich.

KELLY YEARICK is a senior program manager at Forth. She lives and works in Portland, Oregon, where she leads programs to boost EV adoption and access to shared electric transportation. She joined the Forth team in 2017 and has managed various federal and state contracts, including Forth's portfolio of programming as Backstop Aggregator in Oregon's Clean Fuels Program. In her work, she brings together diverse stakeholder groups such as utilities, government agencies, community-based organizations, and industry players to design and implement impactful programs. She currently manages the DOE-funded CRuSE project, a rural carsharing program, and brings her experience around workplace charging to the Electric Vehicle Adoption Leadership Program (EVAL). Yearick holds a master's in environmental management and sustainability from Portland State University, a B.S. from the University of North Carolina at Chapel Hill, and sits on the board of Solar Oregon.

YAN (JOANN) ZHOU is a principal analyst and the group leader of the Mobility and Deployment Group at the Energy Systems Division at Argonne National Laboratory. She leads several DOE- and industry-sponsored research projects on energy and emissions impacts analysis of alternative fuel and vehicle technologies, EV charging behavior, infrastructure modeling, and transportation energy burden analysis. Zhou is a member of the Transportation Research Board's Alternative Transportation Fuel and Technologies committee.

D

Workshop Registrants

Rasay Abadilla, NCDOT
Leetha Abazid, TVA
Robert Abbamont, Westchester County Department of Public Works and Transportation
Amir Abdul-Manan, Saudi Aramco
Ajay Kumar Agarwal, JB Knowledge Park, Faridabad (Delhi NCR, Haryana)
Ramesh Agarwal, Washington University in St. Louis
Luis Aguirre-Torres, City of Ithaca
Mohammed Aslam Ahmad, In-Charge Mapping Design Corp.
Qadeer Ahmed, The Ohio State University
Moe Alahmad, University of Nebraska–Lincoln
José F. Albarrán Núñez, UNAM
Adam Albers, WisDOT
Abuanthony Alim, Alim Innovations
Tareq Aljumylee, ABB
Albert Alonso, Reykjavík University
Igor Alvarado, NI
Walter Alvarado
Luis Alvergue, Arcadis
Vimal Amin, Maryland Department of the Environment
Reza Amini, University of Michigan
Zheng An, Georgia Tech
Martin Andersen, New Jersey Symphony Orchestra
William Anderson, MA DPU
Felicia Angelo
Jorge Aponte Gomez, JAG
Karen Apple, City of Phoenix
Zac Appleton, U.S. Environmental Protection Agency (EPA) Region 9
Nader Ardalan, Ardalan Associates
Allison Arnold, E8
Kristel Arnold, Manitoba Hydro
Michael Arquin, KidWind Project
Miguel Arroyo, MBRDNA
Renata Arsenault, Ford Motor Co.
Atul Arya, IHS Markit
Somayeh Asadi, Penn State
Omar Asensio, Georgia Tech
Steven Ashley
Zohaeb Atiq, FirstEnergy
Chris Atkinson, The Ohio State University
S. Azeem, SMAK
Alham Azhar, Central College
Husain Aziz, Kansas State University
Mona Badie, CEC
Arin Baharian, Passing-Gear
Carla Bailo, Center for Automotive Research
Jeff Baker, Honda Canada, Inc.
Noelle Baker, HATCI - Hyundai Motor Group
Deborah Bakker, Hyundai America Technical Center
Mary Balderrama, North County Transit District (NCTD)
Michael Balshaw
Steve Banashek
Alex Banat, Energy New England

Erica Bannerman, Government
Kevin Barajas, LADWP
Vicki Barbur, Kls
Scott Barfield, NERC
David Barker, North American Subaru, Inc.
Chris Barney, Sonoma County Transportation Authority
Donald Baskin, DoT Volpe Transportation Systems Center
Peter Bateman
Van Baxter, Oak Ridge National Laboratory (ORNL)
Joseph Bayer, National Highway Traffic Safety Administration (NHTSA)
Safak Bayram, UNM
Curtis Baysinger, Vanderbilt Medical Center
Greg Bean, University of Houston
Steve Beers, United Autoworkers International Union
Loida Begley, Government
Kevin Bell, Convergence Research, LLC
Joanna Bellamy, ECCC
Nabil Benktib, nbtechnolology
Brianna Benson, Tennessee Department of Transportation
Rod Beresford, Brown University
Bill Berry, Independent Science Consultant
Shelby Best, Region 1 Planning Council
Larry Biccum, Private citizen
Tony Billera, CATES
Rebecca Blatnica, U.S. Department of Transportation (DOT), Volpe Center
Steve Bloom
Narasimha Boddeti, Washington State University
Stacey Boland, Jet Propulsion Laboratory (JPL)
Emmanuel Bolarinwa
James Bong, Johnson Electric
Massimiliano Borasso, Meccanotecnica Umbra SpA
Anjan Bose, Washington State University
Tamer Bou Hassan, Electric Engine
Elie Bou-Zeid, Princeton University
Naila Boules, Nbmotors
Derek Boyd, University of Maryland
Jeremy Brady, Environment and Climate Change Canada
Maricel Braga, Harvard University and Yale University
Erin Braich, Benton-Franklin Council of Governments
John P. Brazzale, Bridge Advisors
Henri-Paul Bronsard
Margot Brown, World Bank
Don Brownlow, U.S. Army
Doug Brune, U.S. EPA Region 7
Julianna Brunini, Terra Alpha Investments
Lee Ann Bryan, lbryan@itta.com
Anthony Buckley, WSDOT
Norman Bucknor, General Motors
Carrie Burchard, Sandia National Laboratories
James Burke, DeepDive Foresight
Richard Burke, Elders Climate Action
Peter Burnim, Stellar Energy Foundation
Frank Burns, BioPrimate
Jay Butera, Climate Solutions Project
John Byrd, Miller/Wenhold Capitol Strategies
Stephanie Cahill
Brad Cain, C3 Human Factors
Amna Cameron, NCDOT
Luis Canales, LGC Growth Advisors
Jeffry Canin, E8 Angels; Element 8 Fund
Bill Canis, Congressional Research Service
Augusto Cannone Falchetto, Aalto University
Kenneth Capron, MicroRail, Inc.
Josh Carman, PlaceWorks
Matthew Carroll, New York Power Authority
Helotonio Carvalho, Federal University of Pernambuco
Peter Cashin, Imperial Mining Group Ltd.
Geovanni Castano, Dominion Energy
Ricardo Castro, UC Merced
Debapriya Chakraborty, University of California, Davis
Babu Chalamala, Sandia National Laboratories
Georgia Chaloulou, Chevron Marine Lubricants
Kamie Champlin, Utah State University
Jack Chang, ITRI
Robert Charette, IEEE Spectrum
Lionel Charles, Fractal
Phil Chataigneau, Imperial Mining Group
Nirmal Chaudhary, Federal Government
Aden Chen, Taiwan
Boliang Chen, Acbel
Elsa Chen, University of Alberta
Jay Chen, NHTSA
Pingen Chen, Tennessee Technological University
Zhinan Chen, California-China Climate Institute
Lisa Chiladakis, California Air Resources Board
Shih-Miao Chin, ORNL
Christopher Chinapoo, Five Star Quality and Justice Associates Jamaica Ltd
Young-June Choi, JAIN Technology
Jordan Christman, American Petroleum Institute
Jean Chu, Toyota Motor North America R&D

Shadi Chuang, Quanta Technology
Jorge Cisneros, Texas A&M
Terence Clas, Empire State Development - NYSTAR Div.
Bentley Clinton, U.S. DOT
Nan Cochran, Geisel School of Medicine
Ned Codd, WSP USA
Thomas Coffee, Microsoft Research
Treena Colby, Tacoma Public Utilities
Debra Coleman, Coleman Engineering
Carlene Colvin-Garcia, Clean Mobility Consultancy LLC
Daniel Comeaux, CMAP
Lynda Conway, Delta Development Group, Inc.
Mitchell Cook, JustMap Foundation
Richard Corey, California Air Resources Board
Vince Costanzo, Aramco Americas
Gregory Cote, NADA
Chip Cotton, General Electric Global Research
Gerald Couri, House Committee on Energy and Commerce
Marcus Courtney, Courtney Public Affairs, LLC
Chris Coyle, TAO
Stephen Creager, Clemson University
Noel Crisostomo, California Energy Commission
Fergus Crockett
Laura Crossley, DEP
Hector Cruz, The University of Texas at El Paso
John Cruz, MBRDNA
Baozhi Cui, Ames Lab, Iowa State University
Garth Cummings, Lawrence Livermore National Laboratory (LLNL)
James Cummings, UCF
Colin Cunliff, U.S. Department of Energy (DOE)
Héctor M. Curiel, Computotal
Scott Curra, ONRL
Kathleen Curry, Guidehouse
David Cyr, Globe Exports
Ronaldo Da Motta, State University of Rio de Janeiro
Winthrop Dada, TULLEV Innovations
Constance Dadich, DBSA Metro
Jeff Dagle, Pacific Northwest National Laboratory
Qiang Dai, Argonne National Laboratory
Zhen Dai, California Air Resources Board
Brian Daly, CMAP
Vida Daneshmand, Los Angeles Department of Water and Power
Bernard Dardzinski, Blue Collar Science LLC
Bruce Dasinger
Sara Davidson, ITS America - Washington, DC
Stacy Davis, ORNL
Stephen Davis, RV Jensen.com
Damon Day, Bois Forte Tribal Government
Warren Day, U.S. Geological Survey (USGS)
Anibal De Almeida, ISR
Hernane De Almeida, Westchester County DPW&T
Ricardo De Oliveira
Adarsh Deepak, Taksha Institute (TI)
Payman Dehghanian, The George Washington University
Yeenec Delifenroc, KKBPQ
Anthony Della Zanna, Ford Motor Company
Dan Delurey, Vermont Law School
Ahmet Demirbilek, WisDOT
Manoj Desai, ARAI
Debra Desarmeaux, Natural Resource Conservation Service/USDA
Davin Diaz, Benton-Franklin Council of Governments
Crystal Dickerson, NHTSA
Angela Digioia, Chevron
Melvin Din, MIT/School of Secondary and Tertiary Studies Auckland
Cynthia Dinwiddie, Southwest Research Institute
Bradley Distler, Black & Veatch
Deepak Divan, Center for Distributed Energy, Georgia Tech
Donna Dlubac
Jiaojiao Dong, UTK
James Donnelly, Elevate
Caitlin Dorame, GSA
Paul Dravis, Future Perfect Machine
Mike Drummond
Brajesh Dubey, IIT Kharagpur
Tiffany Dubinsky, VDRPT
Larry Dubois, NANOIONIX, LLC
Jason Dunavant
James Dunlavey
Guillaume Dupriez, Aviya Aerospace Systems
David Dvorin, Climate Reality
Mark Eakin, Corals and Climate
Allan Early, Columbia Institute of Renewable Energy Society
Chinedu Edebeatu, Admiralty University of Nigeria, Ibusa, Delta State, Nigeria
Eloise Edom, IÉT
Bill Eger, City of Alexandria
Karen Eggerman, Tensleep Advisory
Shunjiro Eguchi, Microsoft
John Eichberger, Fuels Institute
Kathy Eiler, UC Riverside

Doug Eisinger, Sonoma Technology, University of Washington
Hamed Eksiri, Chevron
Morgan Ellis, Intelligent Transportation Society of America
Jonathan Ells, ERG
Rolando Elvira-Ordonez, LADWP
David Emerling, The Ohio State University
Brian Engle, Amphenol
Miriam Erickson, Brown Williams
Sean Erickson, Western Area Power Administration - Dept of Energy
Wil Escobar, Chevron
Miguel Espitia C., Cordoba University
Beth Etscheid
Gleb Evgenev, MADI
Blaine Fackrell, Morgan County Utah
Martin Faga, The MITRE Corporation
Wei Fan, University of North Carolina at Charlotte
Elaina Farnsworth, The NEXT Ed
Chris Fawson, Utah State University, ASPIRE ERC
Peter Federico, Tufts
Richard Feinberg, Purdue University
Antonio Fernandes, University of Porto, Portugal
Courtney Ferraro, The Cadmus Group, LLC
Ty Ferretti, NREL
Peter Fessenden, Stanford University
Thomas Fincannon, NHTSA
Jeff Firestone, Los Angeles Department of Water and Power
Hannah Fish, NHTSA
Elena Fisher, City of Philadelphia Division of Aviation
Scott Fisher, The NorthBridge Group
John Fleming, Center for Biological Diversity
Kristen Fletcher, Naval Postgraduate School
Lisa Floyd, DOT/NHTSA
Alex Flueck, Illinois Institute of Technology
Jeff Foor, Stellantis
Mike Forster, Mike Forster Consulting
Mickey Fortune, RadTech - The Association for UV & EB Technology
Michael Foster
Hisham Fouad, The World Bank
Guillermo Francia, University of West Florida
C. R. Francis, MSCWB
Kameice Francis, Tufts University
Shelley Francis, EVNoire
Parry Frank, CMAP
Jarrod Franson, Sun Pacific Energy
Doug Fraser, Nature Nanaimo
Dave Freedman, Elders Climate Action
Geri Freedman, Organized Uplifting Resources Startegies
Moses Fridman, Ford Motor Company
Emmanuel Frimpong, SolarTaxi
Alex Fu, FAA
Vivian Fuhrman, Princeton University's Andlinger Center for Energy and the Environment
Anthony Furst, Metro Analytics
Joseph Fustero, North Carolina State University
Washington Gabu
William Gaieck, DOE
Linda Gaines, Argonne National Laboratory
Luis Gamboa, Goodyear Tire & Rubber Company
Anand Gandhi, Vitesco Technologies
Sonia Garakyaraghi, Schneider Electric
Charles Garascia, Consultant
Lorena Garcia, Glenstone
Katya Garcia-Israel, Volpe Center
Reed Garfield, U.S. DOT
Marc Garfinkle, Carbon Point Partners
Michael Garg, Influit Energy
Vijay Garikipati, WL Gore & Associates
Richard Garrett, Sandia National Laboratories
Michael Garrity, MassDOT
Meagan Gates, UNC Chapel Hill
Garrett Gee, House T&I
Isabella Gee, UT Austin
Frank Geefay
Nicholas Georges, Household & Commercial Products Association
Joe Gerber, Connect California
Neil Gerber, Oracle
Maëva Ghonda, Johns Hopkins University
Bob Giannelli, U.S. EPA
Danny Gillich, Manitoba Hydro
Stephen Godwin, Transportation Research Board
David Gohlke, Argonne National Laboratory
Max Gold
Brian Goldstein, EINow.org
Amanda Goller, Colonial Pipeline Company
Andres Gomez, Grand island Area MPO
Sam Goodman, US International Trade Commission
Stefanie Goodwin, National Highway Traffic Safety Administration
Craig Gorin, Dow
James Gorman, GSA
Risei Goto, AP Ventures
Gail Gottehrer, Law Office of Gail Gottehrer, LLC

Varsha Govindu, Ford Motor Company
Mel Graben, BASF
Anthony Granville, World Bank
Kara Grasso, DuPont
Douglas Greenhaus, NADA
Paul Greenough, Saudi Aramco
Stephanie Gresalfi, GSA
Charles Griffith, Ecology Center
Susan Grissom, American Fuel & Petrochemical Manufacturers
Gil Grodzinsky, Georgia EPD
Lana Grom, Student
Perry Gross, Atkins
Lou Guarino
Sebastien Guillon, Hydro-Quebec
CJ Guo, MIT
Matthew Guttenberg, Incepts
Terry Gwinn, Naval Surface Warfare Center
Bahman Habibzadeh, U.S. DOT
Hyeok Hahn, Chevron
Guy Hall, Sacramento Electric Vehicles
Wendy Hamilton, NMSU
Wayne Harasimovitch
Scott Hardman, UCD
Kara Harmet, Region 1 Planning Council
Adrienne Harris, California Air Resources Board
James Harris, Stanford University
Chris Harto, Consumer Reports
Michael Hartrick, Alliance for Automotive Innovation
David Hastings
Ryan Hatch, Pima Association of Governments
Jaap Hatenboer, UMCG Ambulancezorg
Emily Haugh, Guidehouse LLP
Jimmy He, Los Angeles Department of Water and Power
Xin He, Aramco Americas
Brent Heard, National Academies of Sciences, Engineering, and Medicine
Paul Heitmann, NJ BPU Clean Energy
Melanie Henderson, NC-DEQ/DAQ
Chris Hendrickson, Carnegie Mellon University
Darrell Herling, Pacific Northwest National Laboratory
Karlos Anthony Hernandez, Sinnceer Era
Robert Hershey, Robert L. Hershey, P.E.
Chris Hertz, Carnegie Mellon University
Zhuli Hess, William and Flora Hewlett Foundation
Robert Higgins, Higgins Consulting, LLC
Jennifer Hill
Jeffrey Hinkley, NASA
Ethan Hintz, Ryder
Setsuko Hoffman, USDA
Britt Holmen, University of Vermont
K. John Holmes, National Academies of Sciences, Engineering, and Medicine
Kathryn Holmes, ASME
Hyun-Soo Hong, Chevron
Sadman Tamzid Hossain, Centennial College
Colton Hotary, Autos Drive America
Mark Howard, Maryland DOT State Highway Administration
Ken Howden, DOE
John Howes, Redland Energy Group
Brent Hrywkiw, Iconic Power Systems
Hao Huang, University of Houston
Liling Huang, GMU
Yinyan Huang, IAV Automotive Engineering, Inc.
W. Huffstetler, Private
Sharon Huttner, Branford Clean Energy Ad Hoc Committee
Omnia Ibrahim
Takuya Ikeda, Nissan
Tom Ing, Remax
Chris Ionescu, Independent Consultant
Olusiyi Ipingbemi, University of Ibadan
Raphael Isaac, Energetics
Ehsan Sabri Islam, Argonne National Laboratory
Amila J.
Keith Jacobi, State Senate
Michael Jacobson, Johnson Electric
Shyam Jade, Robert Bosch, LLC
Ali Jafarnejad, ODOT
Gary Jakl, SmarTek Edge
Adil Jamal, Duke
Keith Jamison, DOE
Jon Jantz, CE
David Jaskolski, Peach State Truck Centers
Gardner Jason, ORNL
Kathy Jaw, California Air Resources Board
Viktor Jejcic, Agricultural Institute of Slovenia, Dept. of Ag. Engineering and Energy
Megan Jennings, Energy Conservation Works
Phillip Jensen, Chubb Insurance
Leo Yazhou Jiang, Clarkson University
Saraj Jiffry, Jaidah Group
Hyunwoo Jin, ExxonMobil Research and Engineering
Ling Jin, LBNL
Bradford Johnson, HMB Professional Engineers
Evan Johnson, City of Pasadena
James Johnson, The Inetta Fund

John Johnson, Michigan Technological University
Rory Johnston, Langan Engineering, New York, NY
Philip Jones, Alliance for Transportation Electrification
Robert Jordan
Ramond Joseph, NJDCA
Ameya Joshi, Corning
Vishal Joshipura, U.S. DOT
Keith Julien, Johnson Electric
Bryan Just, API
Aditya Kadambi, WCI
Robert Kaplar, Sandia National Laboratories
Ajeya Karajgikar, X2 Fuels and Energy
Som Karamchetty
Dale Kardos, Kardos & Associates, LLC
Farid Katiraei, Quanta Technology
Dimitris Katsoulis, Dow, Inc.
Arthur Katz, OLLI-AU
Gayle Katzman, Westchester County DPW
Richie Kaur, California Air Resources Board
Christopher Kavanagh, NPS
Yue Ke, Argonne National Laboratory
Elise Keddie, California Air Resources Board
Don Kelley, Dakota Rural Action
Nelson Kelly, Macomb Community College
Mohammad Khadhrawi
Fardin Khalili, ERAU
Akasha Khalsa, CA Energy Commission
Ayesha Khan, LUMS University Lahore, Pakistan
Rabia Khan, Washington State University
Robert Kibrick, University of California
Kyung-Hwa Kim, Atlanta Regional Commission
Namdoo Kim, Argonne National Laboratory
Michael King, Colorado Department of Transportation
Sina Ki˙zi˙roğlu, Consultant
Thongchai Klayklueng, Rajamangala University of Technology Isan
Stanley Klein, UC Berkeley
Amy Klinkenberger, Mercedes-Benz Research & Development
Allen Knight, Equitus Financial
Joshua Knight, Pure
Michael Knotek
Peter Kobos, Sandia National Laboratories
Venkat Kolluru, ERM North Atlantic
Eleftheria Kontou, University of Illinois at Urbana-Champaign
Ari Kornelis, Cadmus
John Kosiba, d'Escoto, Inc.
Stephen Kosowski, Kia Motors North America
Kelvin Kotanchek, Evolved Analytics, LLC
Theresa Kotanchek, Evolved Analytics, LLC
Evanthia Kotsi, Kineo
Martin Koubek, NHTSA
Stephen Kraemer, U.S. EPA
Kyle Krall, DNREC
David Kramer, Physics Today
David Krantz, Arizona State University
Philip Krein, University of Illinois at Urbana-Champaign
Ramanan Krishnamoorti, University of Houston
Anand Kumar, Siemens
Shashi Kuppa, NHTSA/DOT
Kenneth Kurani, Institute of Transportation Studies, University of California, Davis
Scott Kuznicki, Modern Traffic Consultants
Haitam Laarabi, LBNL
John Labrecque, IUGG
Grace Lacher, U.S. Green Building Council
Jerry Lacina
Marc Lacrampe, Sterling Strategic Business Development, Inc.
Pallavi Lal, Hatch LTK Engineering Services
Kegan Lam, NA
Susan Lamoreaux, HATCI
Amy Lamson, U.S. EPA
David Landau, Nature's Design
Marc Langweiler
Amanda Larsen, SF Department of Environment
Terry Lathem, Ingevity
Danielle Laurette
Ryan Laws, SCAG
David Lax, API
Dat Le, Cummins, Inc.
Khanh Le, Michelin North America, Inc.
Mr. Le, Montebello High
Kevin Leahy, Grallagh
Yi Ting Leck, Keppel Infrastructure Holdings Pte Ltd.
David Lee, Tennessee DOT
Nathan Lee, UTEP
Robert Leech, Circuit Energy Group
Cr Lei, Ricondo
Tom Lent, Energy & Environmental Transformation
Julieta Anabel Leo Lozano, Centro Mario Molina
Bruno Lequesne, E-Motors Consulting, LLC
Marina Lesse, Naval Postgraduate School
Danny Levy, Influit Energy
Sonya Lewis-Cheatham, VADEQ
Gloria Li, CEQ

Hailin Li, West Virginia University
Vincent Li, TEC
Zuyi Li, Illinois Institute of Technology
Shuang Liang, York Region
Jason Libersky, Quotient Labs
Diana Lidgett, Volvo Cars
Chuck Lieder, SBISD-Technology
Amy Lilly, Mercedes-Benz Research & Development North America
Wongpun Limpaseni, Center for Clean Air Solutions
Marcus Lin, WSP
Chad Lingenfelter, Florida Department of Transportation - District Five
Robin Lisowski, Slipstream
Ev Lit, JRC
Bo Liu, UCLA Luskin Center for Innovation
Hengrui Liu, Tufts University
Yifan Liu, GT
Yilu Liu, UTK/ORNL
Zhu-Jing (Laris) Liu, Bellwether Industries
Chona Liwag, Developlus, Inc.
Sam Lizzio, Johnson Electric Automotive, Inc.
Oscar Llama, Schneider Electric
Nancy Lo, SCAG
Josip Loncaric
Janelle London, Coltura
Douglas Longhitano, Honda
Abdolah Loni, Penn State University
Alisha Lopez, Southeast Florida Clean Cities Coalition
Steve Lorincz, Zoltech
Louie Lozano, LADWP
Zifeng Lu, Argonne National Laboratory
Wison Luangdilok, H2Technology
Tim Luey, St. Catharines Transit Commission
Sergio Luna, The University of Texas at El Paso
Nicholas Lutsey, GM
Janette Lutz, Natural Profit, LLC
Kenneth Lutz, University of Delaware
Robert Lydick, RipTech
W. Lysenko, U.S. DOT
Mingxian Ma, Akron Regional Air Quality Management District
Justin Mabrey, Maryland Department of the Environment
Dimitri Magganas
Andrea Maguire, U.S. EPA
Bryan Maldonado, ORNL
Hassaan Malik
Shiri Mandava
Estelle Mangeney, WMP
Margaret Mann, National Renewable Energy Laboratory
Ellen Mantus, Health Effects Institute
David Manz, Manz Engineering Ltd.
Michelle Markofski, Drone Videography
Antonio Luis Marqués Sierra, University of Oviedo
Allie Marshall, Cadmus
Karen Martin, U.S. EPA
Tanner Martin, Benton-Franklin Council of Governments
Om Prakash Marwal
Alvaro Masias, Ford
George Masinda, Michigan Department of Transportation
Michael Maten, General Motors
Chris Matonti, Chanson Pty Ltd.
Jo Matthews, UAW
Richard Matzner
Lisa Mayberry-Raymond, UAW
Hollis Maye-Key, U.S. EPA
Raul Mb, F4E
Mark McAndrew
Milton McClaren, Simon Fraser University
Joe McCluskey, Burlingame Building Department
Steven McConnell, Marathon Petroleum Company
Ken McCown, University of Arkansas
Hayley McCurdy, Terra Alpha Investments
Gene McHale, FHWA
Jana McKenzie, Logan Simpson
Lucy McKenzie, Atlas Public Policy
Lyle McMillan, Utah Department of Transportation
Sarah Medepalli, Noblis
Michael Medland, Learning Architects
Mohit Mehendale
Jeremy Mehta, DOE
Abdeljalil Mekkaoui, Abdeljalil Mekkaoui DHS
Teo Melis, USGS
David Melko, Placer County Transportation Planning Agency
Tadeu Melo, Petrobras
Joseph Mendelson, Tesla, Inc.
Alexa Méndez, EBC
Sergio Mendoza, Exponent
Aihong Meng, The George Washington University
Rahul Menon, Isani
Meshelle Merritt, Security Building Company
Phil Metz, BioPhysical Economics Institute
Melissa Meuser, CA Air Resources Board
Ana Meuwissen, Robert Bosch, LLC

Javier Meza, IID
Evan Michelson, Alfred P. Sloan Foundation
Brandon Mikkelsen, Environment and Climate Change Canada
Federico Milani, Pegaso
Katie Miller, Adaptive Futures, LLC
Nicholas Miller, Hickoryledge, LLC
Travis Miller, OKI Regional Council of Governments
Camille Minns, Tufts University
Marianne Mintz, Argonne National Laboratory
Negin Moghadasi, UVA
Vladimir Momčllovic´, University of Belgrade - Faculty of Transport and Traffic Engineering
Scott Montgomery, University of Washington
Bryon Moore, WA State Senate
Julia Moravcsik, Coltura
Sawyer Morgan, New Jersey Board of Public Utilities
Kai Morganti, Consultant
Erin Morrow, MWCOG
Joanne Morse, National Academies of Sciences, Engineering, and Medicine
Evvan Morton, AAAS/DOE
Richard Morton
Ronaldo Seroa Motta, State University of Rio de Janeiro (UERJ)
F. H. Mughal
Richard Muisener, Evonik
Samuel Mukasa, University of Minnesota
Partha Mukherjee, Purdue University
Mwiya Mundia, Irish Aid
Matteo Muratori, National Renewable Energy Laboratory
Mark Muriello, IBTTA
Jonathan Murray
Baba I. Musah, Wuhan University of Technology
Amandine Muskus, Kia
Michelle Muzzio, Cell Press
Alyson Myers, Fearlessfund.org
Vinay Nagabhushana, NHTSA
Peter Nagelhout, U.S. EPA
Jan Naidu, Annapurna Solutions
Liz Najman, Recurrent
Yasumi Nakamura-Newbraugh, Nissan North America
Jane Nakano, CSIS
Oliver Nakano-Baker, University of Washington
Yamen Nanne, LADWP
Shravanthi Gopalan Narayanan, Tufts University
Deirdre Nash, New Hampshire DOT
Sindhu Nathan, Stanford University
Lawrence Navin, Methanol Institute
Jason Navota, CMAP
Morteza Nazari-Heris, Pennsylvania State University
Soheil Nazarian, The University of Texas at El Paso
Jordan Neerhof, SparkCharge
Patty Nelis, SLC Airport
Nenad Nenadic, Rochester Institute of Technology
Harrison Neuhaus, Climate Nexus
Kristie Nguyen, PlaceWorks
Ryan Nielsen, Clearway Energy
John Niles, CATES (Center for Advanced Transportation and Energy Solutions)
Jack Nilles, JALA International
Yvette Niwa, Tufts University
Bob Nolty
Thomas Nordstrand, University of Texas at San Antonio
Samy Noureldin, Indiana DOT
Gloria Nuñez, Universidad de Chile
Aterah Nusrat, City of Longmont
Joe Nyangon, SAS
Doug Obey, Inside EPA
Thomas Oconnor, Texas A&M University
Allan Offenberger, University of Alberta
Donna Offutt, Mitch Cox Realtor
Miho Ohsawa
Pat Okita, Upstream Resources, LLC
Charlie Oliver, Tech 2025
Peter Olmsted, FreeWire Technologies
Jan Opper, Opper Strategies & Solutions, LLC
John Orcutt, UCSD/Scripps
Glenn Orveillon, European Commission
Terry Oswalt, Embry-Riddle Aeronautical University
Will Otero, Stellantis
Lee Otis, E8 Clean Tech Angel Investments
Shiqi Ou, ORNL
Gaoyuan Ouyang, Ames Laboratory
Burak Ozpineci, ORNL
Gennet Paauwe, California Air Resources Board
Tom Pacheco, Colorado Department of Transportation
Shane Palmer, Peter Damon Group
Stephanie Palmer, California Air Resources Board
Todd Palmer, Pennsylvania State University
Rakesh Kumar Panda, IIT Kanpur
Sanjeev Pannala, Washington State university
Alyssa Pantaleo, Dartmouth College
Puteri Paramita
Kavita Parikh, BW Engineers and Planners, Inc.
Daniel Parker, UMTRI - Human Factors Group
Max Parness, Toyota

Barry Partridge, Indiana DOT
Ellen Partridge, Shared Use Mobility Center
Avani Patel, Dow
Germana Paterlini, CERTUSOFT - Minneapolis, MN
Ramesh Patro, IIT BHUBANESWAR
Doug Patton, Jupiter Consulting LLC
Morgan Pauley, Indiana Utility Regulatory Commission
Anshul Paunikar, Toyota
Taylor Pavlu, Nevada Division of Environmental Protection
Michael Perella, Crown Castle
Channa Perera, Canadian Electricity Association
Joe Perreira, Nevada Division of Environmental Protection
Thomas Perrot, Energetics
Pamela Person, League of Women Voters Climate Team, Renewable Energy Committee
Andres Pesca, PskEnergy
Eric Petersen, FleetLab
Simon Petrovich, Geely
James Pew, CMAP
Gail Phillips, Cornell University
Rebecca Pickens, University of Michigan
Ralph Pike, Louisiana State University
Billy Pizer, RFF
Jackie Ploch, Texas DOT
Jared Pollock, Sandia
Mike Porcelli, Bronx Community College
Patricio Portillo, Natural Resources Defense Council
Gregory Powell, NHTSA
Joe Powell, ChemePD, LLC
Jim Prall, University of Toronto
Zach Pritchard, DOE
Nika Ptushkina, Machinery Analytics
Sean Puckett, NHTSA
Linda Purdy, North Yorkshire Police
Minahil Qasim, Indiana University
Wenxin Qiao, World Bank
Jim Qin, HJI Group
Yanfeng Qu, Illinois Institute of Technology
Orlando Quezada Simental, The University of Texas at El Paso
Randa Radwan, Highway Safety Research Center (HSRC), UNC
Govind Ragu, MSK
Adeeba Raheem, UTEP
John Rakos, John J. Rakos & Assoc. Consultants Ltd.
Apparao Rao, Clemson University
Parthib Rao, Rice University
Robert Rausch, Transcore ITS, LLC
Carol Lee Rawn, Ceres
Saon Ray, ICRIER
Aleksandar Regent, Polytechnic Rijeka
Griffin Reinecke, Lewis Burke Associates
Benjamin Reinke, X-energy
Weihang Ren, Cu
Carlos Restrepo, New York University
Bob Reuther, White Oak Power, LLC
Spencer Rhee, Vitesco Technologies
Theodore Robin, Jr., PE and Patent Attorney
David Robinson
Reed Robinson, MGC
Sarah Finnie Robinson, Institute for Sustainable Energy BUAmber Robles, GSA Fleet
Emily Rogers, Energy and Environmental Economics, Inc.
Marcy Rood, Argonne National Laboratory
Sarah Rosenwinkel, Arizona State University College of Law
Kris Rosychuk, J.D. Power
Sandy Rotter, Rotter Engineering
Aymeric Rousseau, Argonne National Laboratory
Gustavo Ruffo, Autoevolution
Kyle Ruske, NEXCO-West USA, Inc.
Michael Russell, Michael Russell
Andrew Russo, Illinois EPA
Youngbok Ryu, Northeastern University
Morteza Sabet, Clemson University
Tani Salma, CSIS
Ellery Saluck, Securing America's Future Energy
Balaji Sampathnarayanan, Cummins, Inc.
Juan Manuel Sánchez, Profesor Jubilado de la Facultad de Ingeniería (U.B.A.)
Russell Sanchez, Microsoft
Yuanrui Sang, UTEP
Rohit Sanket, The Ohio State University
José Santiesteban, ExxonMobil
Fabiana Santos, KyaTech Innovation
Kamalen Santos, University of Texas at El Paso
Halim Santoso, Stellantis
Alfred Sarkissian, George Mason University
Isatou Sarr, Medical Research Council Unit, The Gambia at the London School of Hygiene & Tropi
AJ Sauter, University of Colorado Boulder
Troy Savage, Mazzetti
Maxine Savitz, Honeywell
Karma Sawyer, Pacific Northwest National Laboratory
Rebecca Schade, NHTSA

Pete Schihl, U.S. Army Ground Vehicle System Center
Mark Schipper, DOE
Robert Schloss, Schloss Progressive Consulting
David Schmitt, New Jersey Board of Public Utilities
Ann Scholz, NHDOT
Kathleen Schomaker, Town of Hamden, CT
Nick Schou, WRA
Greg Schroeder, Mitsubishi Motors
Pierce Schwalb, Cadmus
Tom Schwerdt, TxDOT
Jeffrey Scoringe, Waka Kotahi. NZ Transport Agency
Norman Scott, Cornell University
Mark Scribner, Energy New England
Prentiss Searles, API
Ted Sears, NREL
Sanjaya Senadheera, Texas Tech University
Tratip Senivongse, Trecon
Murat Senol, University of Strathclyde
Ronald Seubert
Armand Shahbazian, Tacoma Power
Adeeb Shaheer, University of Ottawa
Haziali Shaikh, Blue Ocean Lab
Carolyn Shaw, Actuate
Nasir Sheikh, University of Bridgeport
Lea Sheridan, NJ Transit
Michael Sherman, Ford Motor Company
Meera Shete, Dow Chemical Company
Feifei Shi, Penn State University
Jack Shi, Wanhua
Cassandra Shields, JEPIC-USA
Eric Shierman, Oregon PUC
Duke Shih, Itron, Inc.
Lubna Shoaib, East West Gateway COG
Subhaditya Shom, University of Nebraska-Lincoln
David Shuey, OKI Regional Council of Governments
Anna Shulman, NYPA
David Siewert, LADWP
Philippe Simonis, EPS
J. Simpkins
Patrick Simpkins, AMI Technologies, LLC
Sergey Sinelnikov, National Safety Council
Rajendra Singh, World Bank
Robyn Sinquefield, Federal Transit Administration
Paul Skalny, Fortress 17, LLC
Maraea Skeen, JCDREAM (State of Washington)
Lily Skou, Autos Drive America
Brian Sloboda, NRECA
Alex Smith, Texas Tech University
Ann Smith, UAB
Brett Smith, Center for Automotive Research
Gary Smith, North Carolina Interfaith Power and Light
Amy Snelling, ICF
Stuart Soled, ExxonMobil
Xubin Song, HERI
Abhi Soni, UC
Abhishek Soni, Tesla
Joe Sootsman, Dow
Éric Souque, Manitoba Hydro
Alexander Spathis, Houdini Phantom Research
Dan Sperling, UC Davis
Josh Sperling, NREL
Ryan Spray, Exponent
Siddharth Sridhar, Pacific Northwest National Laboratory
Jill Stark, FHWA
Richard Steeves, rethinkingnuclear.org
Susan Steffenhagen, TDOT
Paul Steier, AAMVA
Edward Stewart, Autos Drive America
Andrew Stillwell, University of Illinois at Urbana-Champaign
Nathan Stockman, Flint Hills Resources
Deidre Strand
William Strand, Cook Children's Physician Network
Michael Stricker, MD Department of the Environment
Andrea Strzelec, UW Madison
Richard Stuebi, Future Energy Advisors
Vivek Sujan, ORNL
Adam Sulkowski, Magna Powertrain
Patricia Sullivan, New Mexico State University
Liang Sun, New Mexico State University
Wuthipong Suponthana, Leonics Co., Ltd.
Shankar Suppiah, BHS KINETIC
Subbarao Surampudi, JPL
Megan Susman, U.S. EPA
David Swanberg, SwanTechnology
John Swanson
Donald Sweezy, NYSDOT
David Sweig, Assemble LLC
Ellen Szymanski, FW Consulting, LLC
Atefeh Taheri, Chevron
Ghazaleh Tahmasebi, Iran University of Science and Technology
Gil Tal, University of California Davis
Hadley Tallackson, Energy Innovation
Don Tan, Northrop
Nathanael Tan, r/FPTO
David Tancabel, EPA

Harry Tang, Johnson Electric
George Tao, APTEC
H Tasaico, NC Department of Transportation
Whitney Tatem, NHTSA
Aleksandr Tea, Bygone
Christopher Tennant, Coordinating Research Council, Inc.
Michael Tess, U.S. Army DEVCOM GVSC
Deborah Thompson, Volvo Group North America
Roy Thompson, Met.
Juan Tibaquira, Universidad Tecnológica de Pereira
Laxman Timilsina, Clemson University
Elena Timofeeva, Influit Enegy, LLC
Cesar Tirado, The University of Texas at El Paso
Daniel Toner, WGBH
T. Michael Toole, University of Toledo
Vernon Torney, Vettra Co.
Alissa Torres, Orange County (FL) Transportation Planning
Loraine Torres Castro, Sandia National Laboratories
Michael Totten, AssetsforLife.net
John Tracy, Tracy-Hayden Associates
Anh Tran, Sandia National Laboratories
Elizabeth Traut, Pennsylvania State University
Roger Trick
Bill Trippe, IEEE
Colby Tucker, U.S. EPA
Dylan Tucker, National Association of State Energy Officials
Van-Rito Ullij, NA
Karen Utt, TVA
Dante Valle, VAS Servicios y Consultoría SAC
Julie Vallese, NHTSA
Alexander Van Boom
Allison Van Pelt, Des Moines Area MPO
John Vang, PlaceWorks
Matt Vanslyke, Energetics
Ana Maria Vargas, U.S. DOT/Volpe National Transportation Systems Center
David Varner, Tekojchoe
Niek Veraart, Michael Baker International
Kirsten Verclas, NASEO
Leonard Vialpando, Broward County EPGMD
Ram Vijayagopal, Argonne National Laboratory
Alvaro Villagran, Shared-Use Mobility Center
Fabian Villalobos, RAND Corporation
Gokul Vishwanathan, Propane Education & Research Council
Christopher Voigt, Virginia DOT
Phil Von Hake, Colorado Department of Public Health and Environment
James Von Suskil, U.S. Navy
Ann Vroom, VroomConsults
Jonathan Wade, University of North Carolina System
Peter Wadhams, Cambridge University
Laurent Waessa, Québec Government Office
Neil Waldick, St. Catharines Transit Commission
Akhil Wali, Morgan State University
Curtis Walker, National Center for Atmospheric Research
Richard Wallace, HWA Analytics
Stephanie Walters, NHTSA
Jack Wang, GAO
Lusha Wang, Washington State University
Michael Wang, Argonne National Laboratory
Renguang Wang, China Automotive Technology and Research Center CO. Ltd
Ross Wang, ORNL
Sumin Wang, E3
Yan Wang, UF
Alan Warde, NYSDOT
Vanessa Warheit, EV Charging Access for All
Dustin Watson, Washington State Dept of Ecology
Barbara Watts, University of Rhode Island
Christopher Weaver, California Air Resources Board
Bernie Weir, Power4LEDs
Andre Welch, Ford Motor Company
David Welsh, SASIPC
Michael Welsh, University of Iowa
Thomas Wenning, ORNL
Tom Wenzel, LBNL
Kira West, TNO
Doug Weston, NHTSA
Josh Weston, NHTSA
Brian Wheeler, Deloitte
Brittany Whited, Department of Energy & Environment
Kevin Whited, City of Carmel
Michelle Wilber, ACEP
Andrew Wilcox, U.S. State Department
Brian Wilkie, National Grid
Brett Williams, Center for Sustainable Energy (CSE)
William Wilms, Advance Renewables
Anitra Wilson, Management and Technology Solutions Group, LLC
Jay Wilson
Kaila Wilson, Vernon Energy District
Linda Wilson, NYS OAG
Kiara Winans, DEQ
Carol Wise

Daniel Wise, Patient Services, Inc.
Timothy R. Wojan, National Science Foundation
Scott Wolfe, TDOT
Susan Wolfe, UAW International Union MI
Candace Wong, University of California, Berkeley
Hon Xing Wong, Center on Global Energy Policy, Columbia University
Michael Woodman, Maryland Department of the Environment
Jeffrey Wuttke, Stellantis
Steve Xiao, Savannah River National Laboratory
Bunli Yang, E4
Mazen Yassine, CUA
Joseph Yetter
Melih Yildiz, University of Kyrenia
Wu Ying, AECOM
Arthur Yip, NREL
Seungju Yoon, California Air Resources Board
Abigail Young, Gensler
Meng Yue, BNL
Bob Yuhnke, Policy committee, Elders Climate Action
Hernani Yulinawati, Universitas Trisakti
Dmitry Zakharyevich, Chelyabinsk State University
Andrew Zalesak, University of North Carolina at Chapel Hill
Thomas Zambrano, AURA Network Systems, Inc.
Mike Zanillo, Citizens Climate Lobby
Samer Zawaydeh, Independent Engineer
Eric Zeise, Eastman Kodak Co.
Elizabeth Zeitler, National Academies of Sciences, Engineering, and Medicine
Hanrui Zhang, Pennsylvania State University
Yun Zhong, Wood Mackenzie
Zhangxin Zhou, Texas A&M University
Bizhong Zhu, Dow Silicones Corporation
Joyce Zhu, American Water
Jun Zhuang, University at Buffalo
Seiar Zia, NHTSA
Kristofer Zimmerman, Marstel-Day, LLC
美萱 肖, University of Oxford

E

Acronyms and Abbreviations

ANL	Argonne National Laboratory
ASME	American Society of Mechanical Engineers
ASU	Arizona State University
BEV	battery electric vehicle
BMS	battery management system
CAFE	Corporate Average Fuel Economy
CAR	Center for Automotive Research
CARB	California Air Research Board
CEC	California Energy Commission
CEO	chief executive officer
DC	direct current
DOE	U.S. Department of Energy
DSCD	Dynamic Systems and Control Division
EO	Executive Order
EPRI	Electric Power Research Institute
EV	electric vehicle
FCEV	fuel cell electric vehicle
GHG	greenhouse gas
ICCT	International Council on Clean Transportation
ICE	internal combustion engine
IEEE	Institute of Electrical and Electronics Engineers

kWh	kilowatt-hour
LADWP	Los Angeles Department of Water and Power
LCA	life-cycle assessment
LDV	light-duty vehicle
LFP	lithium iron phosphate
LIDAR	light detection and ranging
MIT	Massachusetts Institute of Technology
MW	megawatts
NHTSA	National Highway Traffic Safety Administration
NRC	National Research Council
NREL	National Renewable Energy Laboratory
NVES	New Vehicle Experience Study
PEV	plug-in electric vehicle
PHEV	plug-in hybrid electric vehicle
PNNL	Pacific Northwest National Laboratory
RFF	Resources for the Future
SOC	state of charge
SOH	state of health
SUV	sport utility vehicle
UAW	International Union, United Automobile, Aerospace, and Agricultural Implement Workers of America, better known as the United Auto Workers
UC	University of California
UCS	Union of Concerned Scientists
USGS	U.S. Geological Survey
VGI	vehicle-grid integration
VTO	Vehicle Technologies Office
ZEV	zero-emission vehicle